SALVAGE

TAYLOR & FRANCIS *MONOGRAPHS ON PHYSICS*

EDITOR

B. R. COLES, D.Phil., B.Sc.

Professor of Solid State Physics, Imperial College of Science and Technology, London

CONSULTANT EDITOR

SIR NEVILL MOTT, F.R.S.

formerly Cavendish Professor of Experimental Physics, University of Cambridge

MAGNETIC IONS IN METALS

MAGNETIC IONS IN METALS

A review of their study by electron spin resonance

R. H. TAYLOR

Department of Physics, Imperial College, London
and Central Electricity Research Laboratories,
Leatherhead, Surrey, England.

TAYLOR & FRANCIS LTD

LONDON

HALSTED PRESS
division
John Wiley & Sons Inc.
NEW YORK—TORONTO

1977

First published 1977 by Taylor & Francis Ltd, London and Halsted Press (a division of John Wiley & Sons Inc.), New York.

Reprinted from ADVANCES IN PHYSICS, *Volume 24, No. 6, November 1975*

© *1977 Taylor & Francis Ltd.*

Taylor & Francis   ISBN 0 85066 100 5

Printed and bound in Great Britain by Taylor & Francis (Printers) Ltd., Rankine Road, Basingstoke, Hants.

Library of Congress Cataloging in Publication Data

Taylor, R. H.
 Magnetic ions in metals.
 (Taylor & Francis Monographs on Physics.)
 "A Halsted Press book."
 "Reprinted from Advances in Physics, Volume 24, No. 6, November 1975."
 Bibliography: p.
 Includes indexes.
 1. Magnetic ions. 2. Electron paramagnetic resonance. 3. Metals—Magnetic properties.
 I. Title.

QC754.2.M333T39 538'.3 76–53798
ISBN 0–470–99024–4

Preface

This monograph reviews the contribution made to our understanding of the behaviour of localized magnetic moments in metals by the techniques of electron spin resonance. Most of this work has been done in the last ten years and no other comprehensive survey of the literature exists. The main areas of application of the technique are discussed at a fairly simple theoretical level and its limitations are pointed out. Much work remains to be done on the theory of the subject, and the physical interpretation to be given to the parameters derived for the effects of the crystalline electric field, the hyperfine interaction and the effective coupling of local moments to conduction electrons is by no means clear. It is hoped that this survey of the data will assist future theoretical developments. Some suggestions are made in the final section of areas in which electron spin resonance still has a part to play in adding to our understanding of magnetism in metals.

The author would particularly like to thank Professor B. R. Coles whose infectious enthusiasm and valuable advice made this work possible. Dr N. Rivier and Mr S. Male very kindly provided critical reading of the manuscript and useful suggestions. Thanks are also due to many authors of papers from which illustrations have been used for their permission to reproduce them.

October 1976 R. H. TAYLOR

Contents

MAGNETIC IONS
IN METALS

§ 1. Introduction

The study of magnetic impurities in metals by the technique of electron spin resonance is now over 20 years old but to the author's knowledge there exist only two review articles on the subject. The first (Peter *et al.* 1967) was comprehensive at the time of writing but there has since been a sixfold increase in the number of experimental papers published. The second (Orbach *et al.* 1973) provides, in the author's own words, simply " a flavour of the subject which the reader can use to assess its significance and future direction ". The aim of the present article is to provide as comprehensive a survey as possible of the results which have been obtained up to the present time and to give a rather general introduction to some of the theoretical background. It is hoped therefore that the review will satisfy two needs—for a comprehensive directory to the maze of experimental results for the experimentalist or theoretician already working in the field or starting to do so, and for a general assessment of the scope and limitations of E.S.R. as a technique for the study of magnetic moments in metals. An article of this kind will always omit a wide range of experimental results which fall on the boundaries of the area of interest. We have not included, for example, measurements on magnetically ordered alloys where the main point of the work was the study of the ordered state (i.e. ferromagnetic and anti-ferromagnetic resonance) and have also left out the fairly extensive literature on the E.S.R. of the chalcogenides and pnictides of rare-earth elements. Rather more arbitrarily, the results obtained by the technique of transmission electron spin resonance (T.E.S.R.) have been excluded and interested reader is referred in this case to the papers of Monod and Schultz (1968), Monod *et al.* (1973) and Walker (1973).

The histogram in fig. 1 shows the number of papers on the E.S.R. of magnetic ions in metallic hosts published year by year since the work on pure Gd by Kip (A 87) in 1953. The dates of the major developments are indicated along the dateline. The past six or seven years have seen the broadening of the subject to include the observations of resonances from non-S-state ions; furthermore the increasing efficiency of spectrometer systems has allowed impurity concentrations to be reduced to the parts per million range and has made possible the study of alloy systems in which the bottleneck is broken and interaction effects are absent. This increase in spectrometer resolution has also brought the subject to a level where the study of single crystals of dilute alloys has become common-place (nearly all measurements prior to about 1970 were performed on powdered material to give more ions within the skin depth and thus accessible to the microwave radiation).

The article is arranged as follows : in § 2 the main areas to which E.S.R. has contributed are discussed and some theoretical background is given at a simple level. In the short § 3 the severest limitation of the technique, that is the limited

Fig. 1

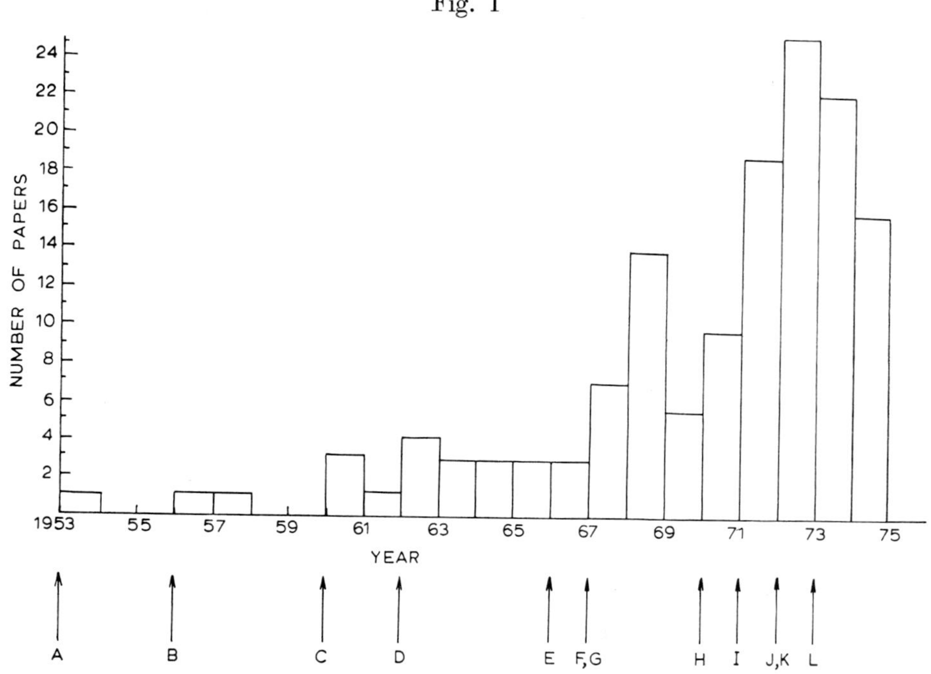

Number of experimental papers on E.S.R. of ions in metals published year by year. Along the timescale, the major E.S.R. advances are shown, A—First observation of E.S.R. in a metal (pure Gd) (A 87). B—E.S.R. of Mn in simple hosts (A 111). C—First observation of E.S.R. in an intermetallic compound (A 113). D—Systematic analysis of Δg and J_{eff}. E—First observation of a non-s-state ion resonance (A 79). F—Experimental investigations of E.S.R. bottleneck. G—Systematic use of single crystals. H—First observation of hyperfine structure (A 26). I—First observation of fine structure (A 24). J—Detection of resonance from a quartet ground state (A 41). K—Investigation of E.S.R. in superconducting hosts. L—Direct observation of excited states (A 57, A 160).

number of ions for which a resonance may be observed in metallic hosts, is examined. In the main section, §4, the results of measurements on alloy systems are systematically reviewed. Finally, in §5, those areas in which E.S.R. still has a contribution to make, and those which appear to be essentially exhausted, are identified. The Appendix contains a full list of references cross-referenced to authors, impurities and hosts and to the various section headings in §2. The body of this article does not cover papers published (or received by the author) after January 1975 but some such papers are listed in the Appendix where they are marked with an asterisk. References in the body of the article are to the list in the Appendix when labelled with a number prefaced by A, and to the list that follows the Appendix when labelled by the year of publication.

§ 2. E.S.R. OF IONS IN METALS—MAIN AREAS OF APPLICATION

The major areas of interest to which the E.S.R. of ions in metals has contributed are the following.

1. The spectroscopic states of ions, their crystal field splittings and fine and hyperfine structures, in various hosts.
2. The E.S.R. bottleneck effect; the determination of conduction electron scattering cross sections and of various relaxation times.
3. The exchange interaction between the localized magnetic moments of ions and the conduction electrons, and the spin polarization of the conduction electron gas; to a limited extent, study of the Kondo effect.
4. Short and long-range magnetic order, including spin glasses and spiral phases, and excitations in spin glasses.
5. The effects of magnetic impurities on type II superconductors.

The principal quantities directly determined by an E.S.R. experiment are the g-value for the magnetic ions and the linewidth of the resonance. Both of these can be studied as functions of the temperature and of the concentration of ions added to the host material.

Measurement of the g-value for dilute alloys first establishes the spectroscopic state of the ion. Indeed for some rare earth ions (Ce, Eu or Yb), the occupancy of the localized f-shell can take two values, only one of which leads to the observation of a resonance (§ 2.4). For ions with $L \neq 0$, measurement of the g-value (strictly, g-tensor) will give information about the role of the crystalline electric field (CEF) in raising the $(2J+1)$-fold degeneracy of the ground state and this important application of E.S.R. is discussed in § 2.5. Although the CEF only operates in first order upon ions with non-zero values of orbital angular momentum, the effect of the CEF on ions such as Gd (4f⁷) and Eu (4f⁷) can lead to fine structure in their spectra as a result of the admixture of higher-energy configurations containing $J = 0$ terms into the ground state when a magnetic field is applied. These effects are of course small, perhaps at least two orders of magnitude less than typical CEF splittings in non-S-state ions, but have been the subject of considerable recent experimental and theoretical interest. The nature of fine structure in the E.S.R. spectrum is discussed in § 2.6, followed by a section dealing with the observation of hyperfine structure and its interpretation.

Unlike the study of magnetic ions in insulators, that of ions in metallic hosts requires consideration of the effects of conduction electrons. The presence of conduction electrons leads to a small shift, Δg, from the g-value observed for the isolated ion, and using a simple molecular field model we can see how this comes about. If we suppose that an ion with a magnetic moment arising from an incomplete d or f-shell is dissolved in a conduction-electron sea of Pauli susceptibility χ_s, if λ is the molecular field coefficient for the interaction between local moments and conduction electrons, M_s the magnetization of the conduction electrons and M_d the magnetization of the local moments at a specified field and temperature, we may write

$$M_s = \chi_s(H + \lambda M_d + \ldots)$$

and the field acting on the d-shell is

$$H_d = H + \lambda M_s + \ldots,$$

i.e.

$$H_d = H(1 + \lambda \chi_s + \lambda^2 \chi_s \chi_d + \ldots).$$

To first order†, and substituting for λ and χ_s in terms of $J(0)$, the local moment-conduction electron exchange integral for $q = 0$, and of $\rho(\epsilon_F)$, the density of states at the Fermi surface, one obtains

$$\Delta g = J(0)\rho(\epsilon_F). \tag{1}$$

The quantity $J(0)$ is not simple and contains a number of contributions. E.S.R. turns out to be perhaps the best technique for the measurement and interpretation of this quantity. The role of E.S.R. in this area is examined in § 2.1.

The conduction electrons normally relax rapidly to the lattice and this leads to a temperature-dependent contribution to the E.S.R. linewidth which is given by the ' Korringa relation ' in the simplest treatment‡ :

$$\frac{d\Delta H}{dT} = \frac{k\pi}{g_s\mu_B}(\Delta g)^2. \tag{2}$$

According to this equation measurement of $d\Delta H/dT$ would add no information beyond that obtained from the g-shift, but the assumptions leading to it require closer examination. In particular, for the linewidth, the response of the conduction electron gas may not be dominated by $J(q=0)$ and exchange enhancement of the conduction electron susceptibility may also play a role. These questions, too, are considered in § 2.1.

Exchange enhancement in the conduction band can profoundly affect the spatial variation of the conduction electron polarization (RKKY polarization); the role of the study of E.S.R. in elucidating the extent of this modification is outlined in § 2.2.

A sufficient treatment of the E.S.R. of ions in metals must in fact include full consideration of the rates at which relaxation may proceed between the *three* components of the system—the ions, conduction electrons and lattice. In some circumstances the relative values of these rates can be such as to modify the measured values of Δg as well as that of $d\Delta H/dT$. This so-called bottleneck can thus complicate the problem of measuring $J(0)$ but, as a compensation, can allow direct measurements to be made of various relaxation rates and scattering cross-sections which would otherwise be difficult to estimate. Section 2.3 treats this matter.

One outstanding problem in relation to the single impurity in contact with a conduction electron sea is, of course, that of the Kondo effect. Potentially E.S.R. could be a useful tool in this area but in practice progress has been disappointing. Some of the reasons for this and the limited extent to which E.S.R. has contributed, will be briefly described in § 2.10.

---

† Higher-order shifts which would be spin and concentration dependent have not been observed experimentally. Yosida (1957) argued that the shift would not be observable since the conduction electron polarization by the d-shell follows the magnetization of the d electron spin adiabatically and the effective field $\lambda^2\chi_s M_d(t)$ cannot produce a torque on $M_d(t)$.

‡ The slope of the linewidth as a function of temperature has been written in differential form in eqn. (2). Frequently, the temperature dependence is expressed in the form $\Delta H = A + BT$ where A is a temperature independent residual width at $T = 0$. Thus $(d\Delta H/dT \equiv B$ and these forms will be used interchangeably in this review.

In alloys with even a moderate concentration of magnetic impurity, interactions between the ions, mediated by the conduction electrons, must be considered. In the limit, of course, these interactions may be so strong that long-range magnetic order may be established and it has already been stated that the details of ferromagnetic and antiferromagnetic resonance will not be considered in this review. Before this stage, however, the role of interactions on E.S.R. behaviour can be considerable and we will consider this in § 2.8.

The study of type II superconductors by doping them with a low concentration of magnetic impurity has provided a potentially exciting new application of E.S.R. An outline of this area of application is given in § 2.9.

2.1. *The E.S.R. g-shift and thermal line broadening. Information about J_{eff}*

One area of considerable interest in which E.S.R. has made a valuable contribution is in elucidating the sign and magnitude of the exchange interaction between localized f or d moments and the conduction electrons. The interaction may be written in the form (Kasuya 1956)

$$H_{\text{ex}} = -2J\mathbf{S}.\mathbf{s},$$

where $\mathbf{S}$ is the spin of the local moment and $\mathbf{s}$ the spin of the conduction electrons. This model describes the physical situation in the limit where the impurity is well localized and where the direct interaction between impurities is small. If the conduction electron wave functions are represented by plane waves and the interaction potential is Coulombic, the exchange integral J may be written

$$J(K, K') = \frac{1}{V} \int \psi^*_{\text{loc}}(\mathbf{r}_1) \exp(i\mathbf{K}.\mathbf{r}_2) \frac{e^2}{r_{12}} \psi_{\text{loc}}(\mathbf{r}_2) \exp(-i\mathbf{K}'.\mathbf{r}_1) \, d\mathbf{r}_1 \, d\mathbf{r}_2. \qquad (3)$$

Usually a number of further approximations are made. To allow for the fact that the Coulomb interaction is reduced by screening effects, the potential is represented by a delta function. In this approximation $J(K, K')$ depends only on the magnitude of $(\mathbf{K} - \mathbf{K}') = q$. For rare-earth ions the total angular momentum J is a good quantum number and de Gennes (1958) suggested that the effects of the orbital moment may be included in the equation for the interaction above replacing $\mathbf{S}$ by its projection on $\mathbf{J}$, i.e. by $(g-1)\mathbf{J}$, where g is the Lande factor. Liu (1961) has examined in greater detail the conditions under which this approximation is valid.

This exchange integral is always positive in sign. However, the total exchange coefficient, J_{eff}, measured experimentally, contains other contributions which in many cases lead to an overall negative coupling. The admixture term in the Anderson (1961) Hamiltonian is one possible source of a negative contribution. Schrieffer and Wolff (1966), and de Gennes (1962), showed that in the magnetic limit the mixing term could be written in the form

$$J_{\text{mix}} \simeq \frac{|V_{Kd}|^2 U}{E_d(E_d + U)}, \qquad (4)$$

where V_{Kd} is the matrix element associated with the s–d interaction between band states and localized states (assumed d-like in this case), U is the Coulomb energy between electrons and E_d is the energy of the centre of the localized state measured from the Fermi energy.

Watson *et al.* (1965) considered the contribution from this mechanism for a Gd^{3+} ion interacting with orthogonalized plane waves. They argued that the dominant source of interband mixing involves the spherical p potential matrix element connecting the 4f shell of Gd and the f-like component of the conduction band. The covalent mixing term should thus be greatest for a matrix with maximum f-character at the Fermi surface. Their calculations suggest that for rare earths the mechanism may indeed be large enough to outweigh the direct exchange contribution and give rise to a negative total J. One of the early successes of E.S.R. in metals was to reveal that negative J's seem to be associated with rare earths in strongly d-like hosts, and this led Shaltiel *et al.* (A 140) to suggest a modification of this scheme whereby the d–f mixing rather than s–f mixing plays the dominant role in contributing a negative component to J. Table 1 shows the signs and values of the g-shift of Gd in a number of hosts (both elemental and intermetallic compounds) and the correlation between a strong negative g-shift and a high density of states in the host d-band is apparent. In this connection should be mentioned the work of Cottet and Peter (A 32) where the measurements of the g-value of Gd in Pd doped with Ni, Rh, Ag showed the role of the d-band density of states particularly clearly.

Whilst these effects may contribute to the large negative g-shifts observed in some systems, Coles *et al.* (A 28) have argued that for Gd, where the $4f^7$ configuration is extremely stable and the energy denominators in the treatment of interband mixing (Watson *et al.* 1965) may be $\sim 10\,eV$ (Dimmock *et al.* 1966), such contributions may play a minor role compared to the effects of d–s hybridization *in the host*. If the d-band susceptibility is high, as in exchange enhanced hosts, and hybridization leads to a polarization of the s-band antiparallel to the magnetization of the d-band, exchange of positive sign between the s-electrons and the 4f shell, may lead to large negative g-shifts. Some evidence is provided for this view by neutron diffraction experiments in pure transition metals (Shull and Yamada 1962) and in the Pd–Fe alloy system (Phillips 1965). At least two other mechanisms have also been proposed. Campbell (1972) argues that the f electron spin on the rare-earth impurity may create a positive local d-moment through f–d exchange and that d–d interactions with other d-moments are then important as in the 3d transition metals. The discussion essentially treats the rare-earth 5d electrons as equivalent to those of a first half transition metal and uses the idea that alloys of transition metals in the same half of the transition series give a positive d–d coupling whilst those from opposite halves give an antiferromagnetic coupling. Davidov *et al.* (A 42, A 53) prefer a model where the g-shifts originate from an antiferromagnetic direct exchange between rare earths and the tight-binding d-shell of the neighbouring host atom. It is to be hoped that the relative importance of these various mechanisms will be elucidated by further polarized neutron studies (of, for example, Pd–Tb) and possibly by N.M.R. measurements, although in this case care must be taken in analysing the complicating effects of core polarization.

The exchange between local moments and conduction electrons is also reflected in a number of other experimental properties of alloys, but comparisons of the values of J obtained by the various techniques should be regarded with caution. The g-shift is related to $J(q=0)$ (eqn. (1)) as is the increase or decrease of the saturation moment from its free ion value revealed by magnetization

Table 1.   Summary of E.S.R. data on intermetallic compounds

Ordering temperature	Compound	g_{pm}	$\dfrac{d\Delta H}{dT}$	T_g	$T_{\Delta H}$	References
$T_N = 15$ K	$EuAl_2$	$1{\cdot}994 \pm 0{\cdot}003$	—	—	—	(A 85) Jaccarino *et al.*
		$2{\cdot}002 \pm 0{\cdot}005$	$12 \pm 0{\cdot}3$	~ 30	~ 30	(A 151) Taylor, Coles
$T_N = 15$ K	$EuAl_4$	$1{\cdot}995 \pm 0{\cdot}004$	—	—	—	(A 159) Wernick *et al.*
		$2{\cdot}000 \pm 0{\cdot}005$	$0{\cdot}3 \pm 0{\cdot}1$	~ 40	~ 40	(A 151) Taylor, Coles
$T_N < 77$ K	$EuPd$	$2{\cdot}01 \pm 0{\cdot}01$	$0{\cdot}8 \pm 0{\cdot}1$	< 77	< 77	(A 151) Taylor, Coles
$T_c \sim 80$ K	$EuPd_2$	$2{\cdot}01 \pm 0{\cdot}01$	$0{\cdot}6 \pm 0{\cdot}1$	—	~ 105	(A 151) Taylor, Coles
$T_c = 168$ K	$GdAl_2$	$1{\cdot}982 \pm 0{\cdot}003$	—	—	—	(A 85) Jaccarino *et al.*
		$1{\cdot}982 \pm 0{\cdot}003$	—	—	—	(A 44) Davidov, Shaltiel
		$1{\cdot}988 \pm 0{\cdot}003$	20	—	—	(A 56) Davidov *et al.*
		$1{\cdot}984 \pm 0{\cdot}003$	—	—	—	(A 80) Hacker *et al.*
		$1{\cdot}989 \pm 0{\cdot}005$	$2{\cdot}1 \pm 0{\cdot}2$	~ 200	~ 200	(A 151) Taylor, Coles
$T_N = 150$ K	$GdAg$	$1{\cdot}982 \pm 0{\cdot}004$	$1{\cdot}6$	~ 200	~ 175	(A 19) Burzo *et al.*
		$1{\cdot}984$	2	—	—	(A 157) Weimann *et al.*
$T_N = 33$ K	$GdAg_3$	$2{\cdot}00 \pm 0{\cdot}02$	$0{\cdot}7 \pm 0{\cdot}2$	60	70	(A 151) Taylor, Coles
$T_N = 31$ K	$GdAu$	$2{\cdot}028 \pm 0{\cdot}01$	14	—	—	(A 14) Burzo, Baican
$T_N = 14$ K	GdB_6	$2{\cdot}01 \pm 0{\cdot}01$	—	—	—	(A 27) Coles *et al.*
		$2{\cdot}00 \pm 0{\cdot}01$	—	~ 60	~ 100	(A 151) Taylor, Coles
		$1{\cdot}992 \pm 0{\cdot}003$	—	—	—	(A 95) Miller, Hacker
		$2{\cdot}00 \pm 0{\cdot}01$	—	~ 60	~ 100	(A 75) Fisk *et al.*
		$1{\cdot}992 \pm 0{\cdot}002$	$0{\cdot}6 \pm 0{\cdot}1$	—	—	(A 143) Sperlich
$T_N = 42$ K	GdB_4	$2{\cdot}10 \pm 0{\cdot}03$	—	—	—	(A 151) Taylor, Coles
$T_N = 140$ K	$GdCu$	$1{\cdot}977 \pm 0{\cdot}004$	$3{\cdot}0$	~ 190	~ 160	(A 19) Burzo *et al.*

Table 1. (*Continued*)

$T_N = 41$ K	GdCu$_2$	$2\cdot013 \pm 0\cdot005$	$2\cdot5 \pm 0\cdot2$	~60	~65	(A 151) Taylor, Coles
$T_N = 18$ K	' GdCu$_4$ '	$2\cdot01 \pm 0\cdot01$	$4\cdot5 \pm 1\cdot5$	~40	~50	(A 151) Taylor, Coles
—	GdCu$_5$	$2\cdot009 \pm 0\cdot007$	—	—	—	(A 140) Shaltiel *et al.*
$T_c = 17$ K	GdCu$_6$	$2\cdot025 \pm 0\cdot01$	7 ± 1	—	~52	(A 151) Taylor, Coles
		—	7	~70	~50	(A 108) Okuda, Date
$T_c = 395$ K	GdCo$_2$	$1\cdot92$	not single valued	~408	404	(A 17) Burzo
$T_c = 90$ K	GdIr$_2$	—	—	—	—	(A 113) Peter, Matthias
		$1\cdot985$	—	—	—	(A 44) Davidov, Shaltiel
$T_N = 86$ K	GdMn$_2$	$1\cdot925$	—	—	—	(A 44) Davidov, Shaltiel
$T_c = 76$ K	GdNi	$1\cdot987 \pm 0\cdot004$	$2\cdot90$	—	—	(A 153) Ursu, Burzo
$T_c = 77$ K	GdNi$_2$	$1\cdot980 \pm 0\cdot004$	$2\cdot45$	—	—	(A 153) Ursu, Burzo
$T_c = 31$ K	GdNi$_5$	$1\cdot944 \pm 0\cdot006$	$6\cdot40$	—	—	(A 153) Ursu, Burzo
		$1\cdot942 \pm 0\cdot007$	—	—	—	(A 140) Shaltiel *et al.*
$T_N = 7\cdot5$ K	GdPd$_3$	$2\cdot04 \pm 0\cdot01$	$6 \pm 0\cdot5$	~11	~11	(A 151) Taylor, Coles
$T_c = 20$ K	GdPt$_2$	$1\cdot8735$	—	—	—	(A 155) Vijaraghavan *et al.*
		$2\cdot032$	—	—	—	(A 44) Davidov, Shaltiel
$T_c = 41$ K	GdPt$_{2\cdot6}$	$2\cdot023 \pm 0\cdot005$	$6\cdot2 \pm 0\cdot5$	—	~60	(A 151) Taylor, Coles
$T_c = 67$ K	GdRh$_2$	$1\cdot967$	—	—	—	(A 44) Davidov, Shaltiel
		$2\cdot005 \pm 0\cdot010$	$3 \pm 0\cdot5$	—	~100	(A 151) Taylor, Coles
$T_c = 68$ K	GdZn$_2$	$2\cdot029 \pm 0\cdot005$	$4\cdot4$	—	—	(A 60) Debray, Ryba
$T_N = 10$ K	Gd$_2$Zn$_{17}$	$1\cdot995 \pm 0\cdot005$	$0\cdot3 \pm 0\cdot1$	~15	~95	(A 151) Taylor, Coles
$T_N = 16$ K	GdZn$_{12}$	$1\cdot980 \pm 0\cdot005$	$0\cdot25 \pm 0\cdot1$	~30	~75	(A 151) Taylor, Coles

measurements. The electrical resistivity, the depression of superconducting transition temperature on addition of magnetic impurities, and the thermal broadening of the E.S.R. linewidth, however, all measure a complicated average over the q-dependence of J across the Fermi surface. This is one of the reasons why the simple Korringa relation (eqn. 2) is rarely obeyed. Inclusion of the effects of exchange enhancement of the host (Narath and Weaver 1968) leads to

$$B = \frac{\kappa\pi}{g_{s}\mu_{B}} (\Delta g)^2 K(\alpha),\tag{5}$$

where

$$K(\alpha) = \left[\frac{(1-\alpha)^2}{[\langle 1 - \alpha F(q)\rangle^2]} \right],\tag{6}$$

$$\alpha = \rho(\epsilon_{F})V.\tag{7}$$

$F(q)$ is the dielectric function, V is an effective interaction potential representing the effects of electronic exchange and correlation within the d-band and the pointed brackets denote an average over all q vectors spanning the Fermi surface.

A more detailed analysis of the q-dependence of the exchange integral and the consequent violation of the Korringa relation was carried out by Davidov *et al.* (A 48). Assuming once again a spherical Fermi surface, the exchange may be written

$$J(\mathbf{K}, \mathbf{K}') \sim J(q) = \sum_{L}(2L+1)P_l(\cos\theta)J^l(K_{F}, K_{F}),\tag{8}$$

where $\mathbf{q} = \mathbf{K} - \mathbf{K}'$ and $|\mathbf{q}| = 2K_{F}(1-\cos\theta)^{1/2}$ and θ is the angle between K and K'. The $J(0)$ entering the expression for the E.S.R. g-shift is given by

$$J(0) = J^{(0)} + 3J^{(1)} + 5J^{(2)} + 7J^{(3)} + \ldots,\tag{9}$$

whilst the average J measured by the thermal broadening of the E.S.R. linewidth may be expressed in the form

$$\langle J^2(q)\rangle = (J^{(0)})^2 + 3(J^{(1)})^2 + 5(J^{(2)})^2 + \ldots\tag{10}$$

By making the further assumptions that each partial wave has a positive contribution from atomic exchange and a negative contribution arising only from interband mixing, J_{mix}, that the negative contribution is maximized for $L=l$, the orbital angular momentum of the electrons and that because of the nearly spherical mixing potential the only other contribution to J_{CM} arises from the $L=l\pm2$ channels, Davidov and his co-workers were able to show that in principle, using similar partial wave expansions for the first Born exchange contribution to the electrical resistivity and the Kondo third-order exchange resistivity, the values of the main partial wave amplitudes for J can be obtained. Values of the amplitudes contributing to the purely atomic exchange part can be obtained from E.S.R. alone, if for a simple s-like host containing a Gd(4f^7) impurity, the partial waves contributing to J_{mix} are totally ignored. Whilst in exceptional cases such an analysis may provide useful estimates of the $J^{(0)}$, for most practical systems the effects of the several approximations must be considerable.

2.2. *Information about the Rudermann–Kittel–Kasuya–Yosida* (*RKKY*) *interaction*

Polarization of the conduction electrons by local moments leads to an indirect coupling between ions ; for two ions of the same ionic species separated by a distance, R, the energy of this interaction (RKKY interaction) may be written

$$E_{\text{RKKY}} = \frac{9\pi Z^2 J_{\text{sf}}^2 (g_J - 1)^2}{\epsilon_{\text{F}}} F(2K_{\text{F}}R)J(J+1), \tag{11}$$

where Z is the number of conduction electrons per atom and, for free conduction electrons,

$$F(2K_{\text{F}}R) = F(\alpha) = \frac{\alpha \cos \alpha - \sin \alpha}{(\alpha)^4} \tag{12}$$

A molecular field model for an ensemble of magnetic ions of the same species gives the following expression for the paramagnetic Curie temperature, θ,

$$k\theta = - \frac{3\pi Z^2 J_{\text{sf}}^2 (g_J - 1)^2}{\epsilon_{\text{F}}} J(J+1) \sum_{i \neq j} F(2K_{\text{F}}R_{ij}). \tag{13}$$

To infer a value for J_{sf} from a measured θ required the evaluation of the lattice sum in this last equation. This has been done for various intermetallic compounds ; Buschow *et al.* (1967) evaluated the sum and measured θ, for compounds of the form $Gd_x R_{1-x} Al_2$, where R is Y, La or Th. The effect of adding Y and La to $GdAl_2$ is to introduce voids in the summation whilst addition of Th (tetravalent) also increases K_{F}. In further papers (Van Diepen *et al.*, 1968, De Wijn *et al.*, 1968) Buschow and his co-workers examined the dependence of the exchange integral on K_{F} for a number of rare-earth compounds (GdAl, $GdAl_3$, and GdCu) using N.M.R. and susceptibility measurements to obtain consistent values of J and K_{F}. A change in sign of J was observed in the vicinity of $K_{\text{F}} = 1 \cdot 3$ Å which it was claimed, agreed with the prediction of Watson *et al.* (1965) that interband mixing becomes the dominant contribution to J at large K_{F}.

Attempts have frequently been made to relate the values of J obtained from E.S.R. studies (see § 2.1) to those obtained in this way from measured values of θ. This type of analysis however treats the problem from a free-electron point of view and ignores effects due to Brillouin zones which cause the Fermi surface to become far from spherical. Stewart (1973) has pointed out this very clearly in emphasizing that the results for J of De Wijn *et al.* (1968) in $GdAl_3$ for example, would imply that the paramagnetic moment should be *reduced* by 28% from the free-ion value, whereas it is in fact, increased by 4%. Furthermore, we note that similar analyses to those described above on $GdZn_2$ by Debray *et al.* (1970) and for $GdNi_5$ by Burzo and Ursu (A 12) do not fall on the curve of J against K_{F} obtained by Buschow *et al.*

It is not widely appreciated that E.S.R. was the first technique to show that the range function of the RKKY interaction differs radically from the form in eqn. (12) above if the host material is exchange enhanced. Kittel (1968) showed that for impurities interacting via an RKKY polarization with the range function in eqn. (12) there should be a broadening of the E.S.R. line on the addition of impurities of a different magnetic species but *not* a shift in the E.S.R. line. This was confirmed in the case of rare-earth impurities in $LaAl_2$–Gd

(A 14).   However, in the strongly exchange enhanced metal Pd, Peter *et al.* (A 114) found that addition of 2% of other rare earths to a Pd 2 at. % Gd solid solution led to both an additional broadening *and* a second-order g-shift.   They showed that the shift was proportional to the added rare-earth's susceptibility and since it was negative for rare earths with an f-shell less than half full and positive in the case of an f-occupancy greater than seven, it was concluded that the interaction between the rare-earth moments and the conduction electrons derives largely from the spin component of the rare earth.   Peter *et al.* assumed that the E.S.R. line shape of Gd in Pd, $P(t)$, was transformed by the addition of other rare earths into

$$h(s) = \int P(t)\, y\,(s-t)\, dt, \tag{14}$$

where $y(s-t)$ is the probability that $P(t)$ is shifted by $(s-t)$ by the action of all the other rare-earth impurities.   They then adopted a simple model, putting $J(r)$ constant for $r < r_0$ and $J(r) = 0$ for $r > r_0$ (in this case $y(s)$ is given by an equidistant set of δ-functions whose strengths follow a Poisson distribution) and found a ratio of shift to width, Z, given by

$$Z = \left(\frac{4\pi}{3}\, C r_0^{\,3} \right)^{1/2} = n^{1/2}, \tag{15}$$

where C is the number of rare-earth ions per cm^3 and n the number within a sphere of interaction around a given Gd.   From their data they obtained $Z = 3$, $n = 9$, $r_0 = 15\,\text{Å}$.   The conclusion is that the RKKY polarization is essentially uniform over several hundred lattice sites.   This was directly and elegantly confirmed by the polarized neutron-diffraction experiments of Low (1964).

2.3. *The E.S.R. bottleneck*

The simplicity and directness of measuring $J(q=0)$ from the E.S.R. g-shift is clouded by the phenomenon of the E.S.R. bottleneck.   Hasegawa (1959) studied the problem of why g-shifts are not observed in the E.S.R. of Mn in Cu when the d–s mixing is probably $\sim \frac{1}{2}\,\text{eV}$, which should lead to a huge g-shift and thermal broadening.   He showed that the reason for absence of a g-shift was related to the amount of cross relaxation between the local moments, conduction electrons and lattice (see fig. 2).

Fig. 2

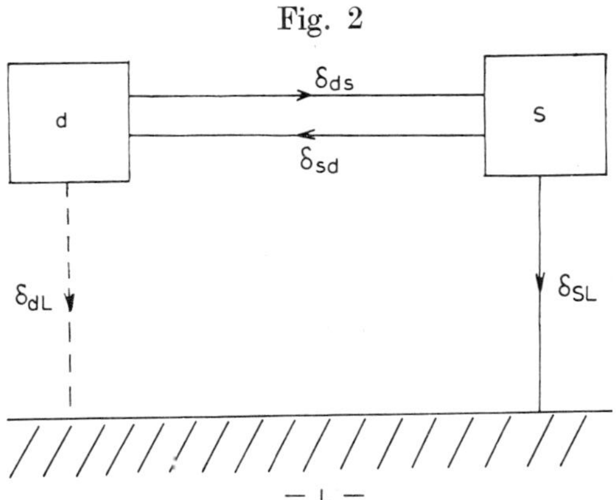

The E.S.R. bottleneck; diagram showing the various relaxation paths.

12   R. H. Taylor

Solving coupled phenomenological equations of the Bloch form, one for the local moment system relaxing to the conduction electrons at a rate $\delta_{sd}(=1/T_{sd})$ and one for the conduction electron system relaxing to the lattice at a rate $\delta_{SL}(=1/T_{SL})$, and assuming relaxation towards the instantaneous internal field and equal g-values for the two spin species, leads to the following equations for the observed g-shift and linewidth as a function of the relaxation rates (Peter $et\ al.$ (1967), Giovannini (1967) Salamon (1971) Schmidt (A 133))

$$\Delta g = \Delta g_0 \frac{(\delta_{SL})^2}{(\lambda\gamma\chi_d H)^2 + (\delta_{SL} + \delta_{sd})^2},\tag{16}$$

$$\Delta H = \Delta H_0 \frac{\left[(\delta_{sd}{}^2 + (\lambda\gamma\chi_d H)^2\left(\dfrac{\delta_{SL}}{\delta_{sd}}\right) + \delta_{SL}{}^2\right]}{(\lambda\gamma\chi_d H)^2 + (\delta_{SL} + \delta_{sd})^2},\tag{17}$$

where λ is the molecular field constant, γ the gyromagnetic ratio for both spin species, χ_d the static susceptibility of the local moment and H the applied field. The direct relaxation from the local moment spin to the lattice is neglected here; the validity for this assumption and the role of this relaxation path has been discussed by Orbach $et\ al.$ (1973) and Yafet (1973).

The values of δ_{sd} and δ_{ds} are given by the Korringa (1950) and Overhauser (1953) formulae respectively:

$$\delta_{ds} = \frac{4\pi}{h}(\rho(\epsilon_F)J_{sd})^2 kT\tag{18}$$

and

$$\delta_{sd} = \frac{8\pi}{h}(\rho(\epsilon_F))J_{sd}{}^2 S(S+1)C,\tag{19}$$

where C is the concentration of magnetic impurities of spin S. These latter two expressions are valid only for high temperatures and Orbach and Spencer (1968) quote expressions which are valid in the extreme low temperature limit.

A detailed balance law

$$\frac{\chi_s}{\chi_d} = \frac{\delta_{ds}}{\delta_{sd}}\tag{20}$$

also holds, where χ_s is the susceptibility of the conduction electron system.

We may consider the expression obtained above in two limiting cases:

(a) Where $\delta_{sd} \ll \delta_{SL}$ the full g-shift and Korringa line-broadening rates are observed. This limit is frequently realized in practice in alloy systems where the host is strongly d-like (e.g. Mn in Pd, Gd in Pd or Gd in LaNi$_5$).

(b) Where $\delta_{sd} \gg \delta_{SL}$ for s-state ions the full g-shift is not observed, being reduced from this value by a factor of $(\delta_{SL}/\delta_{sd})^2$. The linewidth is given by

$$\gamma\Delta H = \frac{\delta_{ds}}{\delta_{sd}} \cdot \delta_{SL}\tag{21}$$

and since from the formulae above

$$\frac{\delta_{ds}}{\delta_{sd}} = \frac{3}{2}\frac{\rho(\epsilon_F)kT}{S(S+1)C}\tag{22}$$

we note that the linewidth in this limit is inversely proportional to the concentration of the magnetic ion and this provides one useful means of identifying bottlenecking in a solid solution alloy system (see below).

The explanation of the bottleneck can be put in a very simple way. In order to maintain equilibrium with the lattice, the local moment attempts to couple some of its energy to the lattice via the conduction electrons. When the conduction electron–lattice relaxation rate is fast enough, the linewidth is dominated by the rate at which the local moment can transfer energy to the conduction electrons and in this case the unbottlenecked (Korringa) behaviour is observed. If, however, the conduction electron–lattice relaxation rate is low, energy can be thought of as 'piling up' in the conduction–electron spin system and the linewidth is that of the coupled system; one observes a single resonance frequency for both the locked spin species. If we are able to decrease the number of magnetic impurities or add some other impurity to provide an increased conduction electron–lattice relaxation rate, then the bottleneck may be broken.

In summary, we expect to identify a bottlenecked system from the following observations:

1. The g-value for an s-state ion will be close to that of the conduction electrons, i.e. usually $g \sim 2$. Any increase or reduction in the magnetic moment of the impurity measured in the static susceptibility will not be observed in the g-value. For a bottlenecked system any value of $J(0)$ calculated from the g-value will be an underestimate, although of the correct sign if $g_{\mathrm{d}} = g_{\mathrm{s}}$.

2. The slope of the linewidth plotted against temperature deduced from the Korringa relation using a 'reasonable' value of J will be much lower than predicted, allowing whenever possible for the effects of exchange enhancement and the q dependence of J. It should be noted that a partial breaking of a bottlenecked system will always be observed in the linewidth before the g-value, since the former varies as $\delta_{\mathrm{SL}}/\delta_{\mathrm{sd}}$ whilst the latter varies as the same ratio squared.

3. The linewidth at a given temperature, after subtracting out any residual term, will usually be inversely proportional to the concentration of magnetic impurity. In a case where the impurity is itself an efficient flipper of conduction electron spins, this concentration dependence will clearly not be observed.

4. Addition of non-magnetic impurities (especially heavy atoms where the spin–orbit scattering of conduction electrons is large) or (even better) magnetic impurities which are not themselves bottlenecked to the host metal, contributes to break the bottleneck and increases the paramagnetic slope of the linewidth of the bottlenecked system.

This latter procedure has been used to obtain quantitative information about the spin-flip scattering rate of the conduction electrons to the lattice for various second impurities (A 76). If we assume three main causes:

(1) scattering in the host material due to imperfections, including the effect of impurities present in the starting materials (this leads to a rate $\delta_{\mathrm{SL}}^{(1)}$);

(2) scattering due to the Gd ions (leading to a rate $\delta_{\mathrm{SL}}^{(2)}$) and

(3) scattering from the added scatterer (giving a rate $\delta_{\mathrm{SL}}^{(3)}$);

we may write

$$\delta_{\mathrm{SL}}^{(\mathrm{total})} = \delta_{\mathrm{SL}}^{(1)} + \frac{\partial \delta_{\mathrm{SL}}^{(2)}}{\partial C_{\mathrm{Gd}}} C_{\mathrm{Gd}} + \frac{\partial \delta_{\mathrm{SL}}^{(3)}}{\partial C_{\mathrm{imp}}} C_{\mathrm{imp}}, \tag{23}$$

14 R. H. Taylor

where C_{Gd} is the concentration of the Gd (or other bottlenecked s-state ion) and C_{imp} is the concentration of the second impurity. By plotting the experimental values of $(d\Delta H/dT)$ as a function of $1/C_{\mathrm{Gd}}$ (with $C_{\mathrm{imp}}=0$) and as a function of C_{imp} for a fixed Gd concentration $\delta_{\mathrm{SL}}^{(1)}$, $\partial\delta_{\mathrm{SL}}^{(2)}/\partial C_{\mathrm{Gd}}$ and $\partial\delta_{\mathrm{SL}}/\partial C_{\mathrm{imp}}$ can be determined.

The cross section $\sigma_{(a)}$ for a process, a, may then be calculated from the equation (A 11)

$$\sigma_a = \left(\frac{\partial\delta_{\mathrm{SL}}^{(a)}}{\partial C}\right)(N_0 V_{\mathrm{F}})^{-1}, \tag{24}$$

where V_{F} is the Fermi velocity.

Yafet (1968) calculated explicitly the cross sections for relaxation for spin-flip scattering of the conduction electrons due to spin–orbit coupling on the impurity and showed that the cross section in both the magnetic and non-magnetic limit varies as ξ^2, where ξ is the orbit coupling parameter. Since $\xi \propto A^{5/2}$, where A is the atomic number of the impurity element, the efficacy of the scattering of heavy elements such as Pt and Th may be well understood. An expression for the cross section due to the conduction electron-impurity spin exchange interaction has been given by Fischer (1971) and by Davidov et al. (A 47). This rate is related to the rate of depression of the superconducting transition temperature by magnetic impurities. Yafet (1968, 1973) and Asik et al. (1969) have considered the detailed processes by which the relaxation δ_{SL} proceeds.

Hirst et al. (A 84) have examined the situation in which two species of paramagnetic impurity are present when both are bottlenecked. For this case the authors derive two interconnected bottleneck conditions and are able to deduce values for the g-value and spin–lattice relaxation of Cr using this technique.

Finally, there are the so-called dynamic effects in the E.S.R. bottleneck. In the extreme bottlenecked case where $\delta_{\mathrm{sd}} \ll \delta_{\mathrm{SL}}$, the term $(\lambda\gamma\chi_{\mathrm{d}}H)^2$ in the denominator of eqn. (16) can be neglected but if the temperature is sufficiently reduced it might be expected in some systems to become comparable with the term $(\delta_{\mathrm{sd}}+\delta_{\mathrm{SL}})^2$ and lead to a reduction in the g-shift and strong dependence on the applied field. It was suggested that such effects are responsible for the observed behaviour of Gd in $\mathrm{LaNi_5}$ (Davidov and Shaltiel, A 45) but considerable doubt has recently been thrown on this suggestion by Male and Taylor (A 93). Dynamic effects have been searched for but not found in a number of other systems, but they *do* appear to be present in the $\mathrm{Eu}(4\mathrm{f}^7)$ in Yb and Yb–Ca systems (Schmidt et al. A 134).

2.4. *Identification of valence states*

Certain ions, particularly those rare earths with a nearly empty, nearly full or nearly half-full 4f shell can, in certain host metals, lower their energy by losing or gaining a valence electron to take up a more stable configuration. Both Eu and Yb are ions of this type. Trivalent Eu has a $4\mathrm{f}^6$ configuration and is non-magnetic ($J=0$), but in many cases Eu prefers to take up a $4\mathrm{f}^7$ configuration and become divalent, establishing, like Gd ($4\mathrm{f}^7$), a half filled f-shell. In this latter state Eu is an S-state ion and should give an E.S.R. signal.

In the same way Yb has a propensity to fill its f-shell; in doing so, the possibility of observing a resonance is lost. Thus the presence or absence of an E.S.R. signal can serve to confirm a valence state; but it must be emphasized that, for some of the reasons outlined in § 3, it cannot always be stated with confidence that the absence of a resonance precludes the possibility of the ion existing in a magnetic state.

2.5. *Effect of the crystalline field*

In rare-earth ions where $\mathbf{J}$ is a good quantum number, at low temperatures the effect of the crystalline electric field (CEF) is to raise the $(2J+1)$-fold degeneracy and the effects of this crystalline-field splitting may be observed in a number of experimental properties of a given alloy system. In cubic crystals, the state of a rare-earth ion can belong to one of eight possible symmetry species, labelled $\Gamma_1\ldots\ldots\Gamma_8$. The ground-state of ions with integral $\mathbf{J}$ can split into states belonging to the first five of these, whilst ions with half-integral $\mathbf{J}$ (which obey Kramer's theorem) give states belonging to Γ_6 to Γ_8. In Er or Dy, the $J = 15/2$ ground-state multiplet may be split by the CEF into three quartets (Γ_8) and two doublets (Γ_6 and Γ_7) whilst in Yb where $J = 7/2$ the splitting is into one quartet (Γ_8) and two doublets (Γ_6 and Γ_7). In Ce (where resonance has only been recently (A 49) observed in a metallic host), where $J = 5/2$, one quartet (Γ_8) and one doublet (Γ_7) are expected.

The ground state in a particular case and the order of the various levels are determined by the sign and magnitude of the cubic crystalline-field coefficients, B_4 and B_6, in the expression

$$H_{\mathrm{CEF}} = B_4 O_4 + B_6 O_6 = \beta C_4 O_4 + \gamma C_6 O_6, \tag{25}$$

where O_4 and O_6 are the fourth and sixth-degree cubic operators (Hutchings 1964). The coefficients are sometimes re-expressed in terms of the expectation values $\langle r^4 \rangle$ and $\langle r^6 \rangle$ for an ion, writing $B_4 = \beta A_4 \langle r^4 \rangle$ and $B_6 = \gamma A_6 (r^6)$, the factors β and γ being those discussed by Elliott and Stevens (1953, 1962).

Or, again, the cubic CEF is sometimes written

$$H_{\mathrm{CEF}} = Wx \frac{O_4}{F(4)} + (1 - |x|) \frac{O_6}{F(6)}, \tag{26}$$

where $F(4)$ and $F(6)$ are numerical parameters given by Lea *et al.* (1962) in their invaluable tabulation of level orders and wavefunctions, W is proportional to the overall splitting and x is given by

$$\frac{x}{1 - |x|} = \frac{\beta F(4) A_4 \langle r^4 \rangle}{\gamma F(6) A_6 \langle r^6 \rangle}. \tag{27}$$

and is a measure of the ratio of the fourth-order and sixth-order parameters.

The full Hamiltonian for the system will of course include a term containing the external magnetic field, $g_J \mu H \cdot J$, and the conduction electron–local moment exchange term $(g_J - 1) J_{\mathrm{fs}} J \cdot s$, in the normal notation.

The point-charge model for face-centred cubic structure and positive charges on the ligands predicts positive values for both $A_4 \langle r^4 \rangle$ and $A_6 \langle r^6 \rangle$, whereas in most alloy systems studied the ground state is consistent with a negative value of $A_4 \langle r^4 \rangle$ and positive $A_6 \langle r^6 \rangle$. The only clear exceptions to this latter rule appear to be Pd : Er and Pd : Dy where both parameters are negative. Coles and Orbach (see the paper of Williams and Hirst (1969)) have argued that a negative value of $A_4 \langle r^4 \rangle$ might be obtained if the screening effect of the partially filled crystal-field split 5d virtual bound state is considered. The d-character of this state ensures that no contribution can be made to the $A_6 \langle r^6 \rangle$ term. The asphericity of the 5d charge distribution due to its splitting into $d\epsilon$ and $d\gamma$ orbitals, which affects the CEF at the 4f electron sites, depends upon the condition that the 5d virtual bound state CEF splitting is larger than the virtual bound state width. In the cases where $A_6 \langle r^6 \rangle$ is negative as well, Dixon and Dupree (1973) have claimed that the presence of 5f and 4f partial wave components in the conduction electron wave functions may account for this. Davidov $et\ al.$ (A 49) have recently discussed the applicability of the various models in greater detail.

The resonances from doublet levels have been, in the main, observed from samples in powder form. Much recent work, in studying the anisotropic behaviour of the quartet resonances, has been carried out using single crystals. In this case the spin-Hamiltonian (Bleaney 1959) is for a cubic crystal field

$$H = g\mu(H_x S_x + H_y S_y + H_z S_z) + f\beta(H_x S_x^3 + \ldots), \tag{28}$$

giving energy levels

$$W = \pm \mu H \left\{ \left[\frac{1}{4}(5\gamma^2 + 3\delta^2) \pm \gamma \left[\gamma + \frac{\delta^2}{2} \left\{ 9(l^4 + m^4 + n^4) - 3 \right\} \right]^{1/2} \right]^{1/2} \right\}^{1/2}, \tag{29}$$

where $\gamma = g + 7f/4$ and $\delta^2 = fg + \delta f^2/2$ and l, m and n are the direction cosines of the magnetic field with the crystal axes. In practice, the g-values of the most anisotropic lines in the quartet are used, together with eqn. (29) above to determine γ and δ and thus to plot the complete angular spectrum for the resonance fields for each of the four transitions.

Recently, the effect of low-lying CEF excited states on the g-value and thermal broadening has been examined. Davidov $et\ al.$ (A 57) have found that in AuEr, $d\Delta H/dt$ becomes non-linear at a temperature corresponding to some fraction of the separation of the excited state from the ground state. Very recently, a separate resonance (A 160) has been observed from the excited Γ_6 state of Er in Pd. In both of these cases direct estimates may be made of the splitting between the ground state and first excited level.

2.6. *Fine structure*

The study of the fine-structure splittings of Gd in single crystals of cubic Au, Pd and LaSb and hexagonal Y and Mg hosts and of Eu(4f^7) in tetragonal BaAl$_4$ has yielded extremely interesting information.

The spin Hamiltonian for Gd(4f^7) in an $axial$ crystal field (e.g. Y or Mg), assuming an isotropic g-value, is given by (see Abragam and Bleaney 1970)

$$H = g\mu S_z H_0 + \tfrac{1}{2}D\{S_z^2 - \tfrac{1}{3}\mathbf{S}(\mathbf{S}+1)\}(3\cos^2\theta - 1) \tag{30}$$

$$+ \text{non diagonal terms (neglected)},$$

where D is the fine structure parameter.

This leads to a theoretical spectrum of seven fine-structure lines each separated by $D(3\cos^2\theta - 1)$ When $3\cos^2\theta = 1$, i.e. $\sim 55°$, the spectrum is expected to collapse into a single line. The centre of gravity of all the lines is given by the expression for the first moment

$$\langle\omega\rangle = \frac{\Sigma_M C_M(E_M - E_{M-1})}{h\Sigma_M C_M}, \tag{31}$$

where C_M is the transition probability between levels M and $M-1$; with the inclusion of Boltzmann factors, i.e.

$$C_M = (\exp(-E_{M-1}/kT) - \exp(-E_M/kT))[\langle M|S_x|M-1\rangle]^2. \tag{32}$$

From the angular variation of the centre of gravity of unresolved lines, the value of D can be estimated.

A second-moment calculation for the same value of D

$$\langle\Delta\omega^2\rangle = \frac{\Sigma_M C_M(E_M - E_{M-1})^2}{h^2\Sigma_M C_M} - \langle\omega\rangle^2 \tag{33}$$

generates the angular variation of the linewidth. Tao *et al.* (A 147) showed that whilst the angular variation of the linewidth due to unresolved fine structure followed the prediction of eqn. (33) the magnitude of the observed linewidth was much less and it was suggested that this discrepancy is caused by an exchange narrowing of the unresolved fine structure.

This situation has been analysed theoretically by Zimmerman *et al.* (1972), Barnes (1974) and Plefka (1972). In a bottleneck system it is argued that the phase coherence between the conduction electrons and local moments leads to *correlated* spin flips at different impurity sites, causing spatial hopping between the fine-structure lines and hence a narrowing. This is why fine-structure lines have never been observed in an extreme bottleneck system such as Cu–Mn. Indeed, this narrowing may well be an important factor in allowing any observation of resonances in powder samples at all. Even in an unbottlenecked system a narrowing is expected as a result of single site spin-flips (Orbach *et al.* 1973).

In the case of cubic materials, such as Au : Gd, the centre of gravity is isotropic at finite temperatures and thus the g-shift shows no angular variation although the linewidth does (Chock *et al.* A 24). Fine structure was completely resolved for Gd in Pd (Devine *et al.* A 65)† and for cubic symmetry the following transitions are expected:

$$\left.\begin{aligned}
H \pm \tfrac{1}{2} &\leftrightarrow \pm \tfrac{5}{2} = H_0 \mp 20\,(1-5\phi)\,b_4/g\mu_B, \\
H \pm \tfrac{5}{2} &\leftrightarrow \pm \tfrac{3}{2} = H_0 \pm 10\,(1-5\phi)\,b_4/g\mu_B, \\
H \pm \tfrac{3}{2} &\leftrightarrow \pm \tfrac{1}{2} = H_0 \pm 12\,(1-5\phi)\,b_4/g\mu_B, \\
H + \tfrac{1}{2} &\leftrightarrow -\tfrac{1}{2} = H_0,
\end{aligned}\right\} \tag{34}$$

where b_4 is the cubic crystal-field parameter and $\phi = l^2m^2 + m^2n^2 + n^2l^2$, where n, m and l are the direction cosines of the magnetic field relative to the crystal axes. Moret *et al.* (A 98) have argued however that a full analysis of the spectrum using the fine-structure narrowing theories above predict that, contrary to observation, the $\pm\tfrac{1}{2}$ transition line will be broadened and absorbed by exchange relaxation into the neighbouring resolved fine-structure lines by virtue of the large spin-flip

---

† Hardiman *et al.* (unpublished) have recently observed a fully resolved fine structure spectrum for Gd in Pt with a value of b_4 three times larger than that found from Gd in any other cubic pure metal hosts.

matrix elements for $\pm\tfrac{1}{2}\leftrightarrow\pm\tfrac{3}{2}$ transitions. The authors argue that, long-range spin–spin interactions lead to a partial collapse of the Gd spins to the $+\tfrac{1}{2}\leftrightarrow-\tfrac{1}{2}$ position and that this effect gives information on the spatial distribution of the spin–spin field at very low solute concentrations. In Pd–Gd it would appear (see § 4.4) that such interactions are constant within a sphere containing about 200 lattice sites but negligible outside.

2.7. *Hyperfine structure*

If the nucleus has a non-zero magnetic moment, a hyperfine structure is normally observed in the E.S.R. of ions in insulators. The first hyperfine splitting for an ion in a metallic environment was observed by Chui *et al.* (A 26) in Ag–Er. But no structure was observed in single crystals of Mn in Cu (A 75) even though the central line was much narrower than the predicted splitting. Barnes *et al.* (1971) attempted to resolve this problem by adding to the molecular field equations of Cottet *et al.* (A 34) a further term arising from the nuclei. The resulting equations were solved for the simplest case of $I=\tfrac{1}{2}$ and generalized to arbitrary nuclear spin. Three regions of interest were identified.

1. In the unbottlenecked limit hyperfine splittings were predicted with the same character as those of a magnetic ion in a dielectric host.

2. In the bottleneck limit such that $\omega_{\mathrm{HF}}\ll\delta_{\mathrm{ds}}$, where ω_{HF} is the electronic hyperfine splitting frequency, it is predicted that hyperfine structure will be suppressed.

3. In the regime where a bottleneck obtains but where $\omega_{\mathrm{HF}}/\delta_{\mathrm{ds}}\gg 1$, it was predicted that hyperfine structure should be observed but with the width of each hyperfine component reduced from the value which would be observed for the same ion in an insulator by $2I/(2I+1)$.

It was suggested that below $1\,\mathrm{K}$, δ_{ds} may be sufficiently small to enable hyperfine structure to be at least partially resolved in systems such as Cu–Mn.

Study of the hyperfine splitting of Er and Yb in Au allowed Tao *et al.* (A 146) to estimate the contributions to the hyperfine coupling constant, due to atomic exchange J_{at} and covalent mixing J_{cm}. Estimates of the contribution of the 6s electrons led to the conclusion that the ratio $J_{\mathrm{cm}}/J_{\mathrm{at}}$ is about 20 times larger for AuYb than for Au : Er.

2.8. *Magnetic ordering and related subjects*

Electron spin resonance provides a useful tool for the study of the onset of magnetic order, both in solid solution alloys and in intermetallic compounds. Although in this review we are not primarily concerned with measurements on fully ordered magnetic materials, most of the alloys which have been studied by E.S.R., except the most dilute, show at least some tendency towards magnetic order, and it is thus important to consider the implications of ordering on the E.S.R. line.

Below the Curie temperature for a ferromagnetic material, the resonance condition was shown by Kittel (1948) to be

$$\omega_0=\gamma\{[H_z+(N_y+N_y{}^{\mathrm{e}}-N_z)M_z]\times[H_z+(N_x+N_x{}^{\mathrm{e}}-N_z)M_z\}^{1/2}, \qquad (35)$$

when the field is in the direction of the spontaneous magnetization (z-axis). N_x, N_y and N_z are the demagnetizing coefficients relating the internal field H^i to the applied field H, i.e.

$$H_x{}^i = H_x - N_x M_x \tag{36}$$

$N_y{}^e$, $N_x{}^e$ are effective demagnetization factors representing the effects of the anisotropy energy, i.e. the net effects of spin-orbit coupling to the lattice and any anisotropic exchange. These expressions may be considerably simplified for specimens of given shape; in the case of a sphere, for example,

$$\omega_0 = \gamma(H + H_A), \tag{37}$$

where $H_A = 2K/M$ for a uniaxial crystal and K is the anisotropy constant. For a powder the demagnetization factors run over a wide range of values, depending on the shapes of the individual crystallites. The observed E.S.R. line becomes an envelope of the individual shifted lines and broadens, but may usually be followed to the lowest temperatures. It is usually noticed that when the magnetization exceeds a few hundred gauss, though the differential peak heights ratio is normally expected to be $\sim 2 \cdot 55$ for an impurity in a metallic host where the skin depth is much smaller than the particle size, the line becomes increasingly symmetric.

Keffer and Kittel (1952) examined the resonance condition for an antiferromagnetically ordered crystal. At $0\,\mathrm{K}$, when the static magnetic field H is applied parallel to the axis of order, and if H_e is the exchange field (isotropic exchange interactions exert couples on ions in this case) and H_A the anisotropy field, the resonant frequency, ω_0, is given by

$$\omega_0/\gamma = H \pm [(2H_E H_A + H_A{}^2)]^{1/2} \tag{38}$$

When (as is usual for s-state ions) $H_A \ll H_e$,

$$\omega_0/\gamma = H_0 \pm (2H_A H_E)^{1/2} . \tag{39}$$

Typical values of the product $(H_E H_A)^{1/2}$ are $\sim 10^5\,\mathrm{G}$, i.e. $280\,\mathrm{GHz}$, so that below the Néel temperature the line will normally be shifted out of the range of conventional E.S.R. apparatus.

One might expect that typical paramagnetic resonance behaviour would be shown above the Néel temperature, T_N. However, in some antiferromagnetically ordering metallic compounds such as GdB_6 and Gd_2Zn_{17}, line broadening has been reported by Taylor and Coles (A 151) at more than $10\,T_N$. This has been attributed to the persistence of magnetic short-range order in the compounds to these remarkably high temperatures, although the precise reasons for such behaviour is far from clear. Similar effects are often present in the form of premature line broadening in solid solution alloys. In one extremely clear and interesting case the role of limited dimensionality in providing short-range order and a g-shift from $g \sim 2$ to $g \sim 3$ over 100 K may be observed. In this compound, $Zn_{13}Mn$ (Caplin and Dunlop, A 21) the g-shift clearly shows up ordering along chains of Mn atoms, well above the bulk ordering temperature for the material, where interchain ordering occurs. Similarly, E.S.R. has recently been used by Bagguley *et al.* (A 9) to study the spiral ordering in Y–Gd alloys. This work opens out interesting possibilities in using E.S.R. as a tool in studying some of the exotic types of magnetic ordering in rare-earth alloys, subject of course to the severe concentration restrictions in using non-s-state impurities.

Finally, the use of E.S.R. in the study of spin-glasses should be mentioned. These materials are metallic systems in which a moderate concentration $(0 \cdot 1 < C < 20 \text{ at. } \%)$ of randomly sited magnetic impurities interact by means (primarily) of the RKKY interaction to produce a well-defined ground state in which spins are frozen into fixed but random directions but which show no magnetic long-range order even at the lowest temperature (Rivier 1974). The archetypal system of this sort is Cu–Mn, which was first investigated in the classic experiments of Owen *et al.* (A 110). They noticed that the E.S.R. line at $4 \cdot 2$ K was radically changed if the sample had been cooled in an applied field and Griffiths (A 78) later investigated these effects in detail, examining the form of the resonant field, linewidth and signal intensity, as a functions of temperature both in field-cooled and non-field-cooled samples. The results showed that in this spin-glass system, at least, the elementary excitations could not be thought of as a simple spin-flip against an internal field.

2.9. *Superconductivity and E.S.R.*

The first observation of E.S.R. of a local moment in the superconducting state was that of Al'tshuler *et al.*† in 1972 (A 4) but the first observations in which the E.S.R. linewidth and the g-value were followed as a function of temperature were those of Rettori *et al.* for Gd in $LaRu_2$ (A 123) and Engel *et al.* for Gd in $CeRu_2$ and $LaRu_2$ (A 73).

The major problem in observing resonance in a type II superconductor is that the upper critical field should not lie within the field sweep of the E.S.R. line. Thus resonance in superconductors is more favourable at X-band (~ 9 GHz) rather than K or Q-band. The role of these factors and the width of the transition are illustrated in fig. 3 taken from Davidov *et al.* (A 51). It is clear from this that of the three frequently studied systems $CeRu_2$, $LaRu_2$ and $ThRu_2$ it is only the first which can be investigated at pumped He_4 temperatures for all normal frequencies.

In each of these systems characteristic changes in the linewidth, g-value and lineshape have been observed. At first sight the change in Δg between the normal and superconducting states would be the simplest to explain. In the super-conducting state the static susceptibility (and thus the g-value) has been shown by Anderson to be reduced from its normal state values. Anderson demonstrated that in the presence of spin–orbit coupling the spin susceptibility χ^s in the superconducting state does not vanish at $T = 0$ but is given by the relation (Anderson 1959)

$$\frac{\chi^n - \chi^s(0)}{\chi^n} = \frac{2V_F \tau_{so}}{\pi \xi_0} \quad \text{for} \quad \xi_0 > V_F \tau_{so}, \tag{40}$$

where ξ_0 is the coherence length at $H = 0$, V_F the Fermi velocity and τ_{so} the spin–orbit relaxation time. In $LaRu_2$ doped with Gd the g-shift is reduced by

---

† In an early paper Shaltiel *et al.* (A 139) failed to observe a change in g-shift for Gd in $CeRu_2$ at the superconducting transition temperature.

Fig. 3

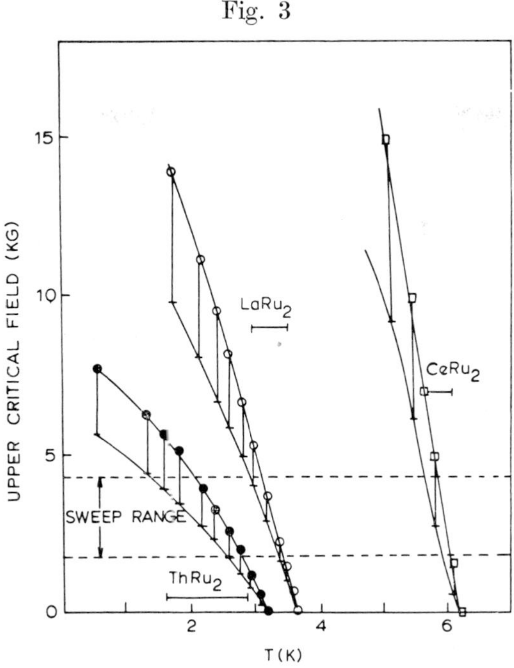

After Davidov *et al.* (A 51)—upper critical field as measured at microwave frequency. The closed circles represent measurements on ThRu$_2$: Gd, the open circles represent measurements on LaRu$_2$: Gd—the rectangles represent CeRu$_2$: Gd (the Gd concentration is 200 p.p.m.). The vertical solid lines represent the width of the transition. The dashed lines are the E.P.R. magnetic field sweep range limits. It is clearly seen that for ThRu$_2$ the sweep range overlaps with the critical-field transition over a wider range of temperature than for LaRu$_2$ or CeRu$_2$. The temperature ranges over which the upper critical field overlaps with the sweep range are shown by horizontal solid lines for the three compounds.

11% in the superconducting state and one is led to believe that this relatively small shift (as pointed out by Rettori *et al.*) constitutes further evidence (to add to the apparent lack of a bottleneck in the normal state) that the spin–orbit scattering is strong in this system. The main surprise in these results was the observation by Engel *et al.* that in CeRu$_{-2}$Gd the g-shift at X-band is in fact increased. The subsequent observation by the same authors, however, that the g-shift *decreases* if measurements are performed in the higher fields at Q-band ($\sim 12\cdot5$ kOe), shows fairly clearly that the sign of the extra shift in the superconducting state is strongly dependent upon the form of the internal field distribution (see later for a detailed account). This important observation allows us some hope that E.S.R. might provide a useful tool for the study of flux distribution in type II superconductors. Although the impurities will disturb such a distribution, the sensitivity of the best spectrometers is now such that only 10 p.p.m. of impurity need be introduced. This possibility is further discussed in §5.

An increase in slope of the linewidth just below T_c and a levelling off again well below T_c as well as interesting changes in the differential lineshape ratio at the transition (from 2·55 in the normal state to a more symmetrical line below T_c) have been observed in each of the superconductors but no complete theory of these effects is yet available. The change in thermal broadening may be analogous to changes in relaxation observed in the N.M.R. of gapless type II superconductors but the theories (see Rettori *et al.* for references) of these were developed for fields immediately below H_{C2}.

Finally we should mention the attempts to correlate E.S.R. properties with the measurements of, for example, re-entrance in the critical field behaviour to obtain parameters such as τ_{so} and the conduction electron exchange scattering τ_{ex}, for various systems, These may, despite considerable approximations in the theoretical estimates, provide a useful method of obtaining order of magnitude estimates of these quantities. In particular comparison of their relative values with the efficacy of the E.S.R. bottleneck in a particular system is of interest.

2.10. *The Kondo effect*

The results of the application of E.S.R. to the study of the Kondo effect have been disappointing.

Spencer and Doniach (1967) showed that for an unbottlenecked system an additional g-shift $\Delta g(2)$ should be observed for a Kondo system and be given by the expression (to second order in J)

$$\frac{\Delta g(2)}{g} \sim (\rho(\epsilon_F)J)^2 \left[1 - \ln \left| \frac{\pi kT}{D} \right| \right],$$

where D is the effective conduction electron bandwidth which appears in the conventional formula for the Kondo temperature (1969), as measured by, for example, electrical resistance measurements (Rizzutto 1974, Grüner 1974). Walker (1968) carried the calculation of the relaxation rate δ_{ds} to third order in J and obtained

$$\delta_{ds} = \frac{\pi kT}{h} \left[J\rho(\epsilon_F) \right]^2 \left[1 + 2J\rho(\epsilon_F) \ln \frac{kT}{D} \right].$$

Kondo terms should thus be measurable in principle, in the temperature dependence of both the E.S.R. g-value and linewidth. Such shifts are extremely small, however, and the contribution to the linewidth is made even more difficult to measure by the large residual widths found for many systems. Criteria for choosing systems likely to yield observable Kondo terms in E.S.R. are that

(*a*) the alloy system must show a clear Kondo effect in its resistivity at accessible temperatures for E.S.R. measurements,

(*b*) the chosen impurity give a well-defined narrow E.S.R. line,

(*c*) the system be not bottlenecked.

One system which would appear to meet these criteria is Yb in Au–Ag alloys and indeed the results of a recent E.S.R. investigation by Nagel *et al.* (A 100) on this system has been interpreted in terms of Kondo contributions to the linewidth. No clear changes in the g-value as a function of temperature were, however, observed.

One point of interest concerns the linewidth of local moment resistances in dilute alloys. Where such moments arise on virtual bound states of 3-d character it might seem at first sight that the E.S.R. relaxation time should be $\sim \hbar/\Delta$, where Δ is the width of the virtual bound state as a resonance in the band structure, since this measures the lifetime of an atomic d-electron on the impurity site. It is clear on reflection, however, and borne out in practice, that the lifetime of the moment and hence of one of its states specified by M_s can be very much greater. On the other hand, the spin-coupling of local moment and conduction electron polarization in the Kondo effect, unlike the formation of a virtual bound state, must set an upper limit of h/kT_K on the local spin lifetime and hence a lower limit of the corresponding paramagnetic resonance linewidth. This argument would seem to forbid the observation of E.P.R. in CuCr ($T_K = 1°K$) but it must be borne in mind that the relaxation time limited to $\hbar/kT_K$ is T_{ds}, and in a strongly bottlenecked system the lifetime of a spin state of the coupled local moment–conduction electron polarization system can be much longer.

For a more thorough discussion of the possible role of the Kondo effect on E.S.R. the reader is referred to the excellent paper of Götze (1973).

§ 3. Observation of a resonance

Perhaps the greatest limitation of E.S.R. in metals as a technique is the fact that resonance may only be observed from a limited number of ionic species. Clearly, only systems in which the impurity in a given host exhibits a 'good' moment need be considered. This rules out, the possibility of observing resonance from 3d transition metals in Al for example. Equally the existence of localized spin fluctuations in dilute solid solutions of (say) Mn in Rh and Pt would also preclude the possibility of doing an E.S.R. study—at least when the characteristic times of such fluctuations are faster than the resonance relaxation time.

A further limitation, although more trivial in nature, should be mentioned and this is the limited solubility of some magnetic elements. We will discuss, for example, many cases of E.S.R. of rare-earth compounds with such elements as Zn and Cu, but using standard techniques of preparation, size factor considerations limit the solid solubility of rare earths in these elements to almost negligible levels.

3.1. *Rare-earth impurities*

The observation of resonance from rare-earth impurities is the simplest case to consider. Once the valence state of the particular element is known we can fairly confidently predict the likelihood of a resonance being observed. Resonance has nearly always been satisfactorily observed for Gd($4f^7$) and Eu in its s-state configuration $4f^7$ in both solid solutions and intermetallic compounds. Among the other rare-earth elements the Kramers ions Dy, Er and (in its $4f^{13}$ configuration) Yb have been observed in a wide variety of hosts. However, due to the strong orbital component of the total moment in these metals the linewidth normally broadens extremely rapidly as a function of both temperature and of concentration and thus it is rare to be able to follow such resonances over an extended temperature regime (say above 10–20 K) and to the best of the author's knowledge no convincing observation has yet been made of resonance from an intermetallic compound in which one of these elements was a partner element. It is perhaps obvious after the remarks of previous sections, but should nonetheless be stressed, that for the many cases in which the CEF ground state of the ion is not a doublet, a single crystal must normally be used.

Of the rare-earth ions other than Eu, Gd, Er, Dy and Yb one extremely tentative (and as the authors themselves pointed out) rather unlikely observation of a resonance from Tm has been reported (A 64). Resonances from Ce in the compounds LaSb and YSb are the only reported cases of a Ce resonance being observed (A 49, A 59). This latter element is particularly interesting—Hund's rules give $J = 5/2$ and this manifold is split by a cubic or octahedral crystal field into a Γ_7 doublet and a Γ_8 quartet, the tables of Lea $et\ al.$ (1962) suggesting that for the latter symmetry a Γ_7 state is lowest whilst for a cubic symmetry a Γ_8 quartet is ground state. However, for Ce the problem is that of finding a host in which Ce is in a well-defined trivalent configuration and where the local moment–conduction electron exchange is relatively weak. The present author has searched for Ce resonances in powder samples of $LaAl_2$–Ce, YAl_2–Ce, Au–Ce, $LaIn_3$–Ce and Rh–Ce without detecting a resonance in the pumped He range. Of these Au–Ce would in many ways offer, it would seem, the most hopeful system of observing a Ce resonance in an elemental host, at sufficiently low temperatures.

3.2. 3d transition metals

This is perhaps the most disappointing area for E.S.R. in metals. $Mn(3d^5)$ has been widely observed in hosts where it exhibits a moment although by no means in all of the cases where a resonance might be expected. The author has failed to observe a $Mn3d^5$ resonance in either Pt or Rh where above fairly small concentrations (a few per cent at most) spin fluctuations are stabilized and other measurements such as electrical resistivity and magnetic susceptibility suggest the existence of good moments in a spin-glass configuration. The likelihood of the observation of a resonance from $Mn3d^5$ is, of course, dependent very largely on the extent of the magnetic resonance bottleneck. The only reason why we are able to observe an E.S.R. line from Mn in Cu at temperatures above a degree or two would seem to be that the strong resonance bottleneck reduces the thermal broadening of the linewidth from a value of perhaps $1000\ T$ to something closer to $8\ T$. In systems such as Pd–Mn where no bottleneck is believed to exist (common to other d-like hosts) the probable reason for observing a resonance here is that the exchange enhancement of the host gives rise to a linewidth reduction of perhaps ten times over the value predicted from the simple Korringa relation on the basis of the observed g-shift. It is perhaps for this reason that Mn in Rh and Pt has not been observed—the bottleneck is not present and the exchange enhancement is not strong enough.

Apart from $Mn(3d^5)$ no resonance from 3d transition metal has ever been observed by the reflection technique (Cr in Cu has been studied by the transmission technique) and the reasons for the non-observation of resonance from other 3d impurities represents an interesting problem. Hirst (1972) has made a notable attempt to understand this problem—he relates bottlenecked E.S.R. to the nature of the degrees of freedom of the magnetic impurity : in particular to the distinction between the spin and orbital components. This he achieves by assigning a definite $3d^n$ configuration, L–S ground state and orbital ground state in the crystal field to the impurity and shows that bottlenecking (and thus observation of a resonance) is only effective when the resonant magnetization of the impurity exchanges flips via the generalized s–d interaction only with the spin magnetization of the conduction electrons. This is the case only for S-state ions or in the case of a degenerate CEF orbital ground state but not first-order spin–orbit coupling (Cr in Cu). In cases such as Fe and Co in Cu where there is a degenerate CEF orbital first-order spin–orbit coupling ground state with bottlenecking is ineffective and no resonance is to be expected. The prediction that Fe in Cu is unbottlenecked seems to be verified by Hirst by studying the results for Fe as a second impurity in Cu–Mn.

Finally, before closing this section we should mention the role of residual linewidth which, if large enough, may preclude observation of an E.S.R. line. It is defined as the $T = 0$ linewidth extrapolated from the Korringa slope (i.e. A in $\Delta H = A + BT$). Analysis of the residual width is an extremely risky enterprise, in that it can arise from many possible mechanisms. We have already mentioned the effects of fine structure (the exchange narrowing of which, we stressed, was perhaps one reason why resonance can be observed at all for magnetic impurities in metals) but unresolved hyperfine structure and inhomogeneous broadening due to interactions make undoubted contributions to the residual width in many cases.

There is also arguably a contribution from the ubiquitous relaxation rate δ_{dL}, the direct relaxation (presumably phonon-induced transitions) from localized moment to lattice but such a contribution has never been experimentally verified.

§ 4. Experimental review

In this section an attempt will be made to describe the main results of the vast number of experimental papers on E.S.R. of ions in metals. In order to inject some semblance of order into this survey, the results are discussed under a number of rather arbitrary headings which classify the impurity as S-state rare-earth ($Gd(4f^7)$ or $Eu(4f^7)$), 3d transition metal or non S-state rare-earth ion in each of a number of types of host material. Dividing the review up in this way at least has the merit of allowing particular measurements to be found with ease and avoids to some degree the repetition which would result from a classification under the conceptual headings of § 2. A link with these categories is made in the directory of E.S.R. results in the Appendix.

4.1. *Gd and Eu metal*

The first observation of paramagnetic resonance in a metal was reported by Kip (A 87) working with a disc of Gd metal. He obtained a paramagnetic g-value of $1{\cdot}95 \pm 0{\cdot}03$ (cf. $g = 1{\cdot}993$ for Gd in a dielectric host). A g-shift was observed to set in at about 50 °C and the line moved continuously to lower fields as the sample was cooled through the Curie temperature (16 °C). Kip found that the half-power half-width also showed an increase from about 400 G at 40 °C to more than 2 kG at 0 °C, below which temperature the line was too broad for further measurement.

Chiba and Nakamura (A 23) extended the work in the paramagnetic regime to 600 °C and found the expected linear relationship between linewidth and temperature above about 350 °C. A temperature-independent g-value of 1·97 was found in the same temperature regime. Popplewell and Tebble (A 118), in their work on the Y–Gd alloy system (discussed below), obtained a value of $g = 1·96 \pm 0·03$ in the paramagnetic regime. Burzo and Domsa (A 16) obtained a paramagnetic g-value of $1·972 \pm 0·006$ and obtained a figure of 4·80 for the Korringa slope. A minimum in the linewidth was observed at ~ 330 K and a g-shift was found to set in just above this temperature. A large negative residual linewidth was obtained for measurements at both X-band and K-band, a slightly higher value being obtained at the higher frequency. The paramagnetic g-value was found to be frequency independent.

Careful work above and below the Curie point using a single crystal led Rodbell and Moore (A 126) to quote a g-value of $1·94 \pm 0·02$ in the paramagnetic regime and a value of $g = 2·00 \pm 0·02$ below T_C.

Europium metal, which is known to order in a complex antiferromagnetic arrangement with a Néel temperature of 94 K, was studied using E.S.R. by Peter and Matthias (A 113). Between room temperature and 120 K they obtained $g = 1·985 \pm 0·015$ and a linewidth of 1300 G. Below 120 K, however, the line was found to 'broaden and disappear'. The 'free-ion' g-value for Eu is normally given as 1·993 in an insulating host.

Ehara and Arrott (A 69) believed that they were able to detect a slight frequency dependence of the g-value of pure Eu, the X-band value of $g = 2·04 \pm 0·01$ being slightly higher than values obtained at higher frequencies. The resonance line was found to shift to lower fields and broaden below 120 K, with an even more rapid broadening setting in below about 90 K. This behaviour was discussed in terms of the possible modes into which Eu may order, but no firm conclusions could be reached using samples in powder form.

4.2. *Rare-earth intermetallic compounds*

Peter and Matthias (A 113) made the first observation of resonance in an intermetallic compound in their work on $GdIr_2$. They failed to observe a resonance in $EuIr_2$, showing that in this compound the Eu is trivalent. This work was extended to the compounds $EuAl_2$ and $GdAl_2$ by Jaccarino et al. (A 85) who studied their E.S.R. behaviour in the temperature regime 100 K to 300 K. Allowing for demagnetization corrections, they obtained values of $\Delta H = 900 \pm 30$ G for the half-power linewidth at room temperature and a g-value of $1·982 \pm 0·003$ in the case of $GdAl_2$ and $g = 1·994 \pm 0·003$ for $EuAl_2$. In corresponding N.M.R. measurements on $(R.E.)Al_2$ alloys, they observed Knight shifts consistent with a negative exchange interaction but which was found to vary in magnitude across the series. This was the first measurement of the sign of the conduction–electron polarization for a magnetic impurity in a metal. They were able to explain neither the inconstancy of the N.M.R. data, nor the sign of the exchange. Peter (A 112), however, showed that the addition of other rare earths to $GdAl_2$ led to a broadening of the E.S.R. line due to RKKY interactions and found that the line broadenings had the same *relative* values as the shifts in the N.M.R. work.

Shaltiel et al. (A 140), in the course of their extensive work on resonance in S-state ions, measured the g-value and linewidth of the two $CaCu_5$ structure compounds, $GdNi_5$ and $GdCu_5$ at 78 K and 65 K respectively. They obtained g-values of $1\cdot942 \pm 0\cdot007$ and $2\cdot009 \pm 0\cdot007$ and linewidths at the above temperatures of 905 ± 90 G and 875 G. They did not investigate the temperature dependence of these parameters. The negative g-shift in the $GdNi_5$ compound was interpreted as being due to the high density of states in the almost full d-band of this compound, whilst the positive shift in $GdCu_5$ was ascribed to the predominance of true atomic exchange.

E.S.R. in the compound $EuAl_4$ was studied by Wernick *et al.* (A 159). Their magnetization measurements at $4\cdot2$ K showed that the compound was meta-magnetic, undergoing a transition to ferromagnetism in a field of about 15 kG. A plot of $1/\chi$ versus T suggests antiferromagnetism below 6 K. A high (295 K) temperature g-value of $1\cdot995 \pm 0\cdot004$, with a half-width of 430 G, was obtained and the resonance was also observed at $1\cdot8$ K, where the line had broadened to over 3 kG. If this line was indeed due to $EuAl_4$ and not Eu oxide (which is frequently found to give a broad line in Eu compounds at pumped He_4 temperatures) the result is surprising—the observation of a resonance below the ordering temperature being characteristic of a ferromagnetic transition rather than of a transition to antiferromagnetism, and leads one to speculate that such behaviour may be related to the metamagnetic properties of the alloy.

Vijaraghavan *et al.* (A 154, A 155) performed N.M.R. experiments on compounds of the form $(R.E.)Pt_2$ and $(R.E.)Pt_5$, where $(R.E.) = Ce$, Pr and Nd. They obtained different Knight shifts from the inequivalent Pt sites in the $(R.E.)Pt_5$ compounds. All the values of J obtained from these measurements were, however, found to be positive and in disagreement with an observed negative g-shift ($g = 1\cdot873 \pm 0\cdot007$). However, Davidov and Shaltiel (A 44) (and later Taylor and Coles (A 151)) contradicted this result, obtaining a g-value of $2\cdot032$. The former authors also measured the g-values of the compounds $GdAl_2$, $GdRh_2$, $GdMn_2$ and $GdIr_2$, finding negative g-shifts for these, the value for $GdAl_2$ agreeing well with the previous measurements (A 85, A 112). In N.M.R. experiments on the same compound they distinguished between two hyperfine fields: the first resulting from the polarization of the conduction band due to the Gd ion of the same nucleus, and the second from the conduction electron polarization arising from the same ions. The second contribution was estimated to be about 15% of the first. Allowing for core polarization, they obtained fairly good agreement between the relative values of the hyperfine field and E.S.R. g-shifts. Hacker *et al.* (A 80) obtained a further value of $g = 1\cdot984 \pm 0\cdot003$ for $GdAl_2$ at room temperature and Davidov *et al.* (A 56) as part of their study of bottlenecking in the $LaAl_2$–Gd system obtained $1\cdot988 \pm 0\cdot008$, both substantiating the small negative g-shift observed in previous measurements.

Okuda and Date (A 108) measured what they presumed to be a dilute solution of Gd in Cu. The form of their results, however, and the fact, in particular, that the data did not change with concentration shows that they did in fact measure an intermetallic compound. Comparison with the data of Taylor and Coles (A 151) shows that this compound was Cu_6Gd. They found a constant g-value from room temperature down to about 60 K (value not quoted) where a shift in the resonance field to lower values sets in, reaching a value of about $2\cdot5$ kG at 10 K. The linewidth data were linear at high temperatures, with a Korringa slope of about 5 G . K^{-1}. A fairly sharp linewidth minimum was found at 50 K, the line broadening to about 2 kG at 10 K.

The temperature dependence of the g-value and linewidth for the compound $GdNi_5$ was studied by Burzo and Ursu (A 12) in the temperature range 100 K to 300 K (i.e. above the ordering temperature ~ 30 K). They obtained a temperature-independent g-value of $1\cdot944 \pm 0\cdot006$ (in good agreement with Shaltiel *et al.* (A 140)) and a Korringa slope of $6\cdot4$ G. K^{-1}. A value of $K(x)$, the exchange enhancement factor in the Korringa relation, of $0\cdot1$ was required to bring the values of Δg and $d\Delta H/dT$ into agreement. The effects of the q-dependence of J however was not considered. In further publications (A 13, A 153) Burzo and his co-workers reported the results of measurements on two further Gd–Ni compounds—GdNi and $GdNi_2$. For these compounds g-values of $1\cdot978 \pm 0\cdot004$ and $1\cdot980 \pm 0\cdot004$ and Korringa slopes of $2\cdot90$ G. K^{-1} and $2\cdot45$ G. K^{-1} were obtained. These values were found to be the same at X-band and K-band within experimental error but the residual linewidths were less at lower frequencies. As might be expected, poor agreement was obtained for J_{s-f} deduced from the E.S.R. measurements and from the RKKY interaction formula using the measured value of the paramagnetic Curie temperature (see § 2.2).

Burzo and his co-workers have also obtained results for the equiatomic Gd–noble metal compounds GdCu, GdAg and GdAu. In the case of the first two of these compounds (A 19) the resonance was followed down to the Néel points (140 K and 150 K respectively) just above which line broadening and a temperature dependence of the g-value became apparent with no signs of the extensive regimes of magnetic short-range order reported for some other antiferromagnetically ordering compounds by Taylor and Coles (A 151). The Korringa slopes ($3\cdot0$ G. K^{-1} and $1\cdot6$ G. K^{-1}) are approximately half the values which would be deduced from the g-values ($1\cdot977 \pm 0\cdot004$ and $1\cdot982 \pm 0\cdot004$). The negative signs of these shifts do not agree with the *enhancement* of the Gd moments obtained in static susceptibility measurements or with the notion that Gd partnered by a simple metal host gives a positive g-shift. However, it is possible that for such high Gd concentrations the d-character of the rare earth may itself be important. We note that the measurements of Davidov *et al.* (A 48) on LaAg, YCu and YAg doped with small quantities of Gd (see below), where the bottleneck was opened, lead to positive shifts. These results would seem to provide evidence for the two-band bottleneck model of Davidov *et al.* (A 56). The compound GdAu (A 14) gave a positive g-shift ($g = 2\cdot028 \pm 0\cdot010$) and a Korringa slope of 14 G. K^{-1}. The difference in sign of the shift between the Cu and Ag compounds and the Au compound is puzzling, but the negative sign of the shift in GdAg was confirmed by Weimann *et al.* (A 156, A 157) who obtained $g = 1\cdot984$ and $d\Delta H/dT = 2$ G. K^{-1}, in excellent agreement with Burzo.

A study of the compound $GdCo_2$ (A 17, A 18) revealed one particularly interesting feature—the linewidth, after passing through a sharp minimum at about 405 K ($T_c = 395$ K) showed a curvature in the high temperature linewidth as a function of temperature, the slope becoming smaller as the temperature was increased (fig. 4). The g-value passed through a weak minimum at roughly the same temperature as that of the linewidth and had not become temperature independent at the highest temperature (~ 425 K) at which measurements were taken. Extrapolation of the data would suggest a slightly negative g-shift at high temperatures.

Fig. 4

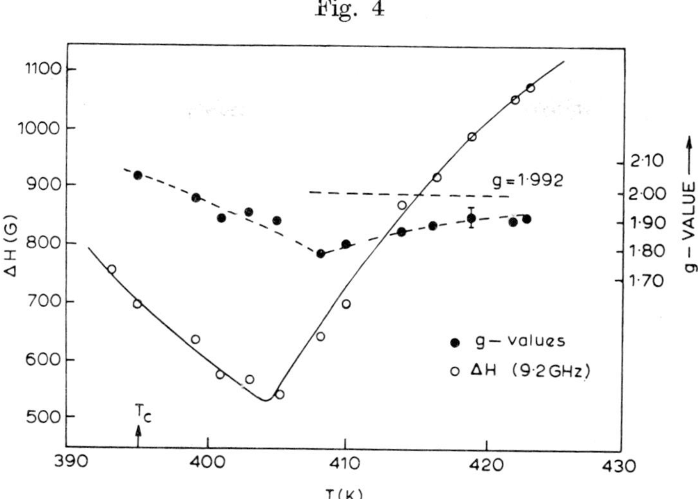

Temperature variation of the linewidth and g-values for $GdCo_2$ compound after Burzo *et al.* (A 17, A 18).

Fig. 5

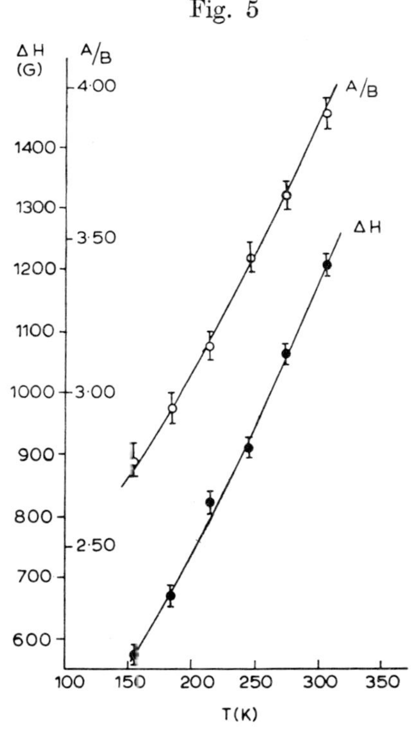

Lineshape (A/B) and linewidth as a function of temperature for $GdZn_2$ (Debray and Ryba (A 60)).

The compound $GdZn_2$ was measured from 150 K to room temperature by Debray and Ryba (A 60). A temperature-independent g-value of $2·029 \pm 0·005$ was obtained and a slope of the linewidth in the paramagnetic regime of

$4 \cdot 4 \, \mathrm{G.K^{-1}}$. The authors observed a rather puzzling temperature dependence of the A/B ratio (fig. 5), which reduced linearly from a value of about $4 \cdot 0$ at $300 \, \mathrm{K}$ to about $2 \cdot 7$ at $150 \, \mathrm{K}$. In theory, a value of $2 \cdot 55$ would have been expected, independent of temperature. A value of $J = + 0 \cdot 018 \, \mathrm{eV}$ was extracted from the g-shift assuming a value of $K(\alpha) = 0 \cdot 1$. This value for $K(\alpha)$ would seem to be somewhat large for Gd partnered by a 'simple' metal host. Comparison was made with the value of J extracted from the RKKY expression for θ.

E.S.R. measurements on metallic rare-earth borides have been made by a number of workers. Coles *et al.* (A 27) found a g-value of $2 \cdot 01$ for GdB_6 at both room temperature and at $77 \, \mathrm{K}$ but no resonance was visible at $4 \cdot 2 \, \mathrm{K}$. They observed a 20% broadening on cooling from room temperature to $77 \, \mathrm{K}$. The non-observation of the resonance at liquid helium temperatures was in accord with the results of their resistivity measurements which indicated antiferromagnetic ordering at $14 \, \mathrm{K}$. Miller and Hacker (A 95) obtained g-values of $1 \cdot 994$, $1 \cdot 992$ and $1 \cdot 992$ at $77 \, \mathrm{K}$, $195 \, \mathrm{K}$ and $295 \, \mathrm{K}$ respectively with half-power widths of 733, 609 and $630 \, \mathrm{G}$ at the same temperatures. These g-values are in good agreement with the value of $2 \cdot 00 \pm 0 \cdot 01$ obtained by Fisk *et al.* (A 74) at both $300 \, \mathrm{K}$ and $77 \, \mathrm{K}$ but the linewidths measured by Miller and Hacker are

Fig. 6

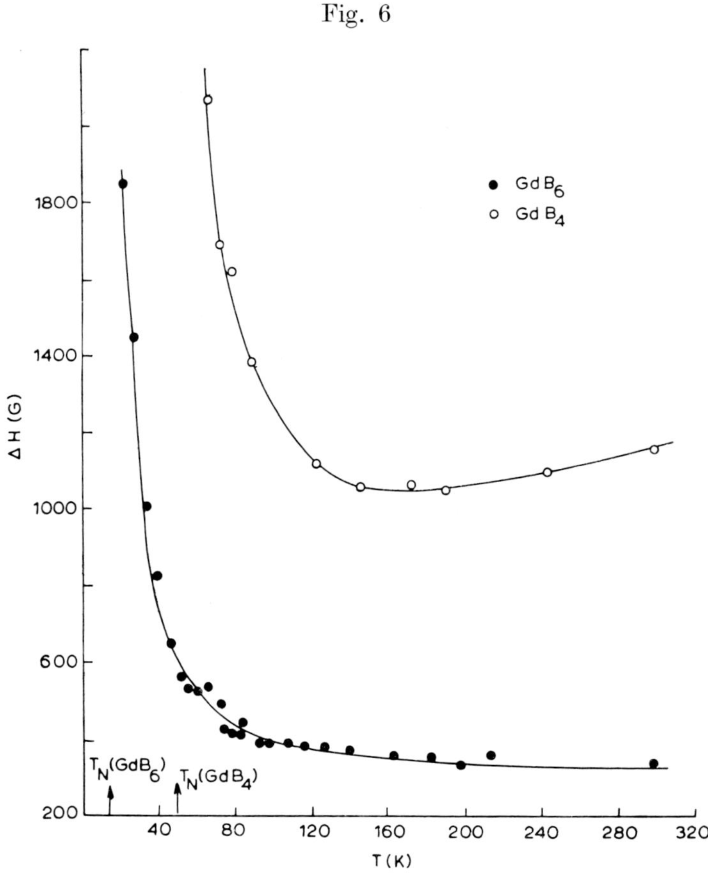

Linewidth as a function of temperature for the compounds GdB_4 and GdB_6 (Fisk *et al.* A 74).

almost twice as large as those obtained by Fisk *et al.* (A 74). In this latter paper the linewidths and g-values of both GdB_6 and GdB_4 ($T = 45$ K) were shown as a function of temperature over the range 4·2 K to 300 K. The linewidth of both alloys were found to broaden on approaching the Néel points (fig. 6), but in both cases the line broadening was found to set in at very high temperatures. In GdB_6 a true Korringa slope had not been achieved at even $20T_N$, whilst the comparable figure for GdB_4 ($g = 2·01 \pm 0·03$ at 300 K) was over $4T_N$. In both cases the Néel temperatures are lower than those expected from a comparison with Tb and Dy compounds, the ratio $|\theta|/T_N$ is high for these compounds and anomalies were detected in both the electrical resistivity and static susceptibility well above T_N. Taylor and Coles ascribed these effects to the existence of a significant amount of short-range order well above the Néel points No comparable effects were observed in the resonant frequencies.

The linewidth of the compound GdB_6 was followed to even higher temperatures (~ 900 K) by Sperlich (A 143), who showed that linearity is in fact only achieved above about 600 K. The frequency independent minimum

Fig. 7

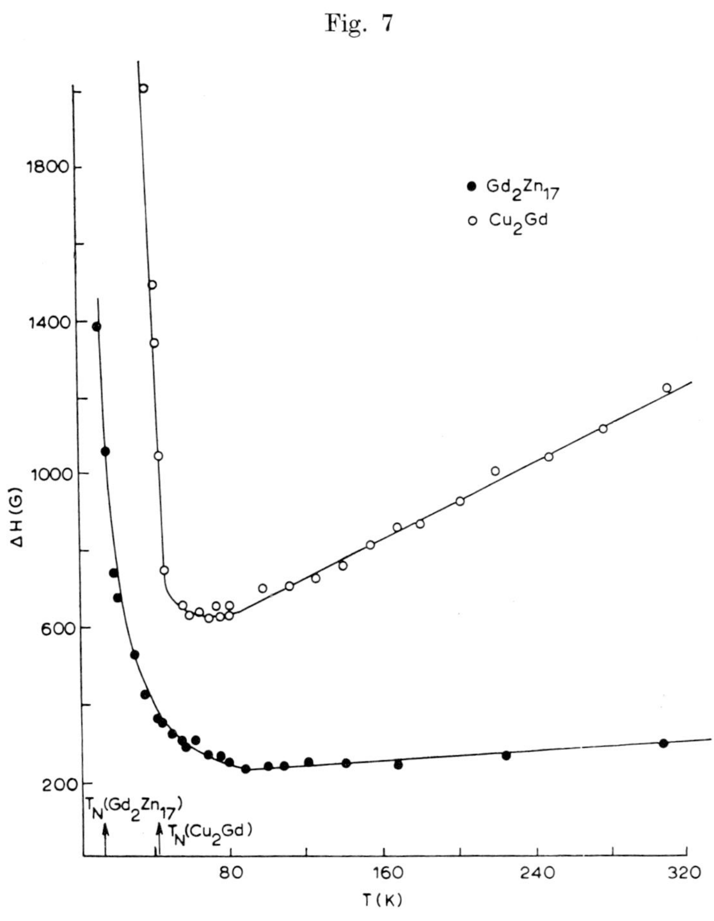

Linewidth as a function of temperature for the antiferromagnetically ordering compounds $GdCu_2$ and Gd_2Zn_{17} (Taylor and Coles, A 151).

linewidth obtained by Sperlich ($\sim 300\,\mathrm{G}$) is less than either of the previously obtained values. The frequency dependence of the linewidth of GdB_6 and the semiconductor EuB_6 was measured by Sperlich and Janneck (A 141). They argued that a dipolar width of $2\cdot33\,\mathrm{kG}$ would be expected for EuB_6 at 300 K whereas a line 767 G broad at X-band and 648 G broad at Q-band is observed. In GdB_6 a dipolar width of $2\cdot43\,\mathrm{kG}$ would be expected. Using the theory of Anderson and Weiss (1953) they were able to explain the frequency dependence of the EuB_6 line and the frequency independence in GdB_6 and estimated values of the exchange fields of $25 \pm 2\,\mathrm{kG}$ and $65 \pm 2\,\mathrm{kG}$ for the two compounds.

A detailed study of the temperature dependence of the linewidth and resonant field for the ferromagnetically ordering compounds $GdAl_2$, '$GdPt_{2.6}$', $GdCu_6$, Pd_2Eu and $GdRh_2$ and for the antiferromagnets $GdAg_3$, $GdCu_2$, $EuAl_2$,

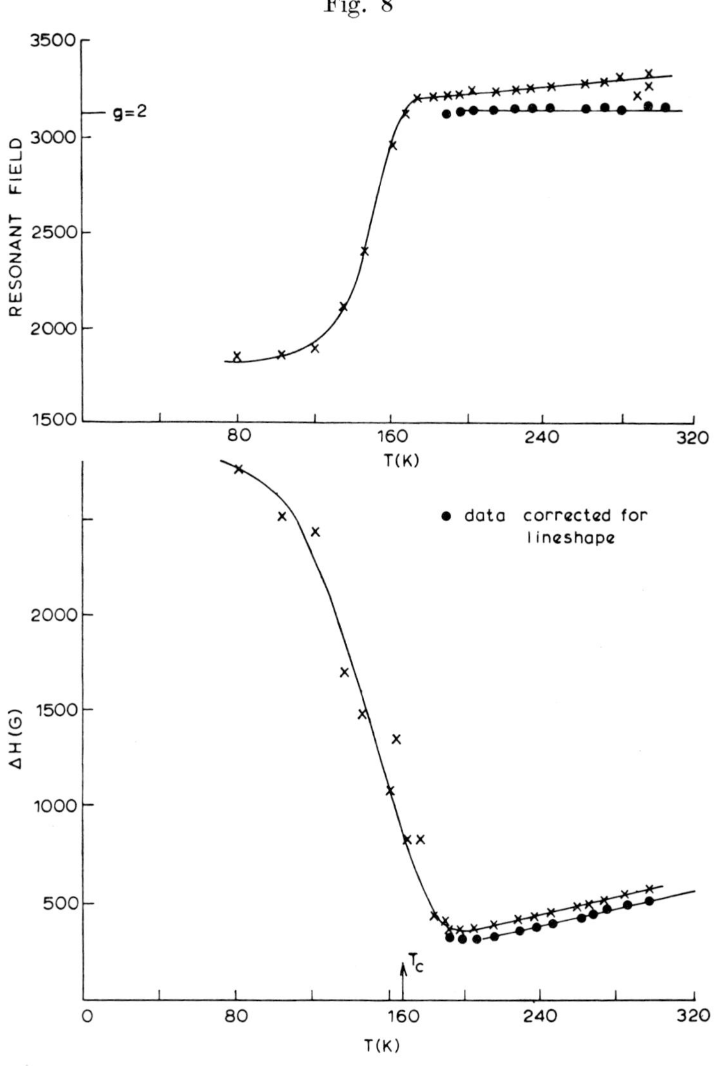

Linewidth and resonant field as a function of temperature for the compound $GdAl_2$ (Taylor and Coles, A 151).

'GdCu$_4$', EuAl$_4$, GdZn$_{12}$ and Gd$_2$Zn$_{17}$ has recently been published by Taylor and Coles (A 148, A 151). They showed that the ferromagnetically ordering compounds show a shift in resonant field which follows, approximately, the form of the magnetization in the X-band resonant field ($\sim 3\cdot2\,$kG). The linewidth of the polycrystalline powders begins to broaden (see fig. 8) at roughly the same temperature that the shift becomes apparent and this temperature is normally $1\cdot3$–$1\cdot5\,T_{\rm c}$. Of the compounds studied, including two Mn compounds Pd$_2$MnSn and 'Zn$_{10}$Mn'—which will be discussed below—only one compound failed to show this typical behaviour. In Cu$_6$Gd the broadening set in at $3\,T_{\rm c}$ and it was thought possible that this anomalous behaviour may be associated with the effects of atomic disorder. Gd–Pt Laves phase compounds with a wide range of stoichiometry (from Pt$_2$Gd to Pt$_3$Gd) showed somewhat similar effects in E.S.R., the ratio of the temperature at which the linewidth begins to shift, $T_{\Delta H}$ to the Curie temperature being much larger for the compounds at the ends of the series than for those (such as 'Pt$_{2\cdot6}$Gd') near the middle.

In the antiferromagnetically ordering compounds, Taylor and Coles showed that in some cases line broadening sets in at or less than $2\,T_{\rm N}$ (Ag$_3$Gd, CuGd and EuAl$_2$) but in others, as in the borides, broadening in the E.S.R. linewidth occurs at up to $10\,T_{\rm N}$ (Zn$_{17}$Gd$_2$). In many cases, where data exist, anomalous behaviour in $1/\chi$ versus T, resistivity and ratios of $\mathrm{l}\theta\mathrm{l}/T_{\rm N}$ have been observed above $T_{\rm N}$. The behaviour of one of the 'anomalous' compounds, Gd$_2$Zn$_{17}$, is compared in fig. 7 with an example of a 'well-behaved' compound (Cu$_2$Gd). This excess line broadening was ascribed again to the presence of an extended regime of short-range order above the Néel point and comparisons were drawn with similar types of E.S.R. behaviour in non-metallic compounds. In these materials the limited dimensionality of the crystal lattice has often been shown to be important in leading to short-range order, but in the metallic compounds studied by Taylor and Coles some of the crystal structures (e.g. GdB$_6$) are so simple that such contributions can be ruled out. The effects, instead, it was suggested, might arise from the conflicting constraints of lattice periodicity and the RKKY interaction.

The compound Pd$_3$Gd, whilst showing clear signs of antiferromagnetism in its susceptibility and resistivity, appeared in its E.S.R. behaviour to resemble a ferromagnet. It was suggested that this property might be common to materials which undergo a metamagnetic transition in fields comparable with the E.S.R. resonant field.

The measured g-shift and line broadenings for these compounds and for all the others discussed in this section, are summarized in table 1. Relevant magnetic and crystallographic data for most of these compounds with full references to previous magnetic measurements are to be found in the review article on rare-earth compounds by Taylor (1971).

4.3. *Intermetallic compound hosts with S-state rare earth impurities*

This section is one of the largest because it embraces most of the subject area headings of §2. Substitution of Gd and Eu in compounds has been particularly useful in studying the magnetic resonance bottleneck and more recently these are the types of alloys used for studying type II superconductors by magnetic resonance.

In their wide-ranging early resonance study, Shaltiel *et al.* (A 140) substituted 5 mole % Gd on the A sites of the AB_5 compounds shown in the table below. All the compounds gave negative g-shifts except $LaPt_5$ and YCu_5, the largest negative shift being obtained for the Ni compounds.

	g	ΔH
$LaNi_5$	$1{\cdot}877 \pm 0{\cdot}007$	$550 \pm 10\%$
YNi_5	$1{\cdot}900 \pm 0{\cdot}007$	$525 \pm 10\%$
$ThNi_5$	$1{\cdot}913 \pm 0{\cdot}007$	$525 \pm 10\%$
UNi_5	$1{\cdot}953 \pm 0{\cdot}007$	$470 \pm 10\%$
$ThIr_5$	$1{\cdot}973 \pm 0{\cdot}007$	$640 \pm 10\%$
YCu_5	$2{\cdot}000 \pm 0{\cdot}007$	$350 \pm 10\%$
$LaPt_5$	$2{\cdot}022 \pm 0{\cdot}007$	$860 \pm 10\%$

It has, of course, become clear since this early work that a 5% substitution of Gd may result in appreciable bottlenecking and that interaction effects may well be present in some compounds even at 20 K. As the signs of the g-shifts in YNi_5 and YCu_5 were opposite, the two compounds were alloyed. At the composition $YNi_{4.5}Cu_{0.5}$ a maximum in the g-shift and a minimum in the linewidth were obtained but, as extra Cu was added, the g-value moved back towards $g = 2$. The addition of other rare earths substituted on the A sites gave an increase in g-shift for elements to the right of Gd in the rare-earth series and a decrease in the g-shift for those to the left (fig. 9).

Fig. 9

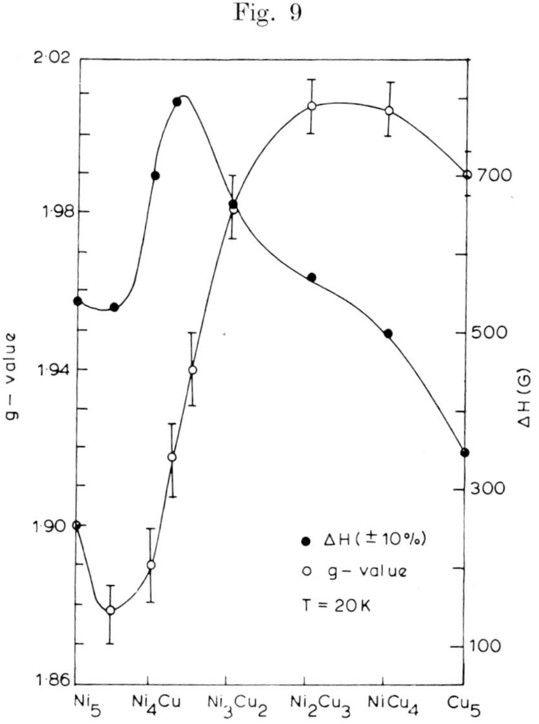

The g-value and linewidth of 5 mole % Gd in YNi_xCu_{5-x}, as a function of Ni–Cu concentration at 20·4 K (Shaltiel *et al.*, A 140).

A plot of linewidth against temperature for 2 at. % Gd dissolved in the compound UPd_3 gave a linear relation between about 60 K and 240 K with a rapid line broadening at low temperatures. It was concluded that this increase in linewidth was due to magnetic ordering. The g-value increased from 2·02 at 300 K to 2·10 at 4·2 K.

A number of important conclusions were drawn from these observations:

1. The similarity of the $(Ni-Cu)_5Y$ g-shift measurement with measurements on the (Pd–Ag) and (Pd–Rh)–Gd systems suggested that $LaNi_5$, YNi_5 and Pd are rather similar hosts.

2. It was observed that when trivalent La and Y in the Ni_5 compounds are replaced by tetravalent Th and hexavalent U, the g-shifts are reduced. This was attributed to filling of the d-bands. It also demonstrates the absence of a resonance bottleneck in these compounds.

3. When Ni was replaced by Ir and Pt, smaller g-shifts were obtained. This decrease was explained by the fact that these elements are the 5d analogues of Co and Ni and show only weakly enhanced paramagnetism in the elemental state.

4. The large negative g-shifts were explained on the basis of the d–f interaction mechanism discussed in §2. The small positive shifts observed for the other compounds were taken as an indication that in these alloys 'true' s–f exchange was stronger than the d–f term.

In a further paper, Shaltiel *et al.* (A 139) made a systematic investigation of the sign and magnitude of the g-shifts of 1–5% Gd substituted into AB_2 Laves phase compounds on the A site where A = Sc, Y, La, Th, U and Zr. The B sites were occupied by 4d and 5d elements with an almost full d-band (Ru, Rh, Ir and Pt). The g-shifts are shown below:

$ZrRu_2$	$-0·02$	$LaRh_2$	$-0·01$
$ScRu_2$	$-0·05$	$LaRh_2$	$--0·02$
YRu_2	$-0·06$	$LaPt_2$	$+0·01$
$CeRu_2$	$-0·03$	$La_{0.15}Th_{0.85}Ru_2$	$-0·07$
$ThRu_2$	$-0·03$	$LaAl_2$	$0·00$
$LaRu_2$	$-0·13$		

When these are plotted as a function of increased filling of the B-element d-band, i.e. Ru/Os, Rh/Ir and Pt, a trend from large negative g-shifts to small positive shifts becomes clear.

Both E.S.R. and N.M.R. experiments (A 137) were carried out by Shaltiel *et al.* on the $La_{1-x}Th_xRu_2$ system with 1 at. % Gd substituted on the La sites. The g-shifts and Knight shifts, K, were found to vary with x at 20 K, both showing a minimum shift at the 50–50 composition. A linear relation was found to exist between both Δg and K, and between K and the susceptibility of the $x = 0$ compound as the temperature was varied. From these measurements a value of $-0·05$ eV was estimated for the effective exchange between the f-moment and the host d-band.

Cottet *et al.* (A 33) investigated compounds of the form $La_{1-x}(R.E.)_xRu_2$ by susceptibility and E.S.R. measurements. On substituting 6% Gd they found that the alloy became ferromagnetic at 9·5 K with $\theta = +16$ K. Addition of Tb and Pr to the Gd alloy increased the susceptibility in the first case and decreased it in the second. E.S.R. on the Gd alloy showed some temperature dependence, a broad minimum being exhibited at about 25 K, where $g = 1 \cdot 88$. Addition of 1% Pr shifted this curve to lower values of g and addition of 1% Tb shifted it to higher values. The increase in linewidth and shift back towards $g = 2$ at low temperatures was attributed to dynamic effects in the E.S.R. bottleneck.

Davidov and Shaltiel (A 38, A 45) looked at compounds of the same form containing 0·5, 1·0, 3·0 and 10% Gd, and the compound $LaNi_5$ with 0·1, 1·0, 2·0 and 5% Gd substituted on the La sites. In the $LaRu_2$ alloys they found that for concentrations of 3·0 and 10% Gd there was, again, considerable temperature dependence of the g-value, their results being broadly similar to those of Cottet *et al.* At low concentrations, they found that Δg was almost temperature independent and equal to $-0 \cdot 15$. On the addition of 1% of Pr to a 0·25% Gd alloy, the additional g-shift varied approximately as $1/T$, indicating that the g-shift was proportional to the Pr susceptibility.

The g-shift results for the $LaNi_5$–Gd alloys are shown in fig. 10.

Fig. 10

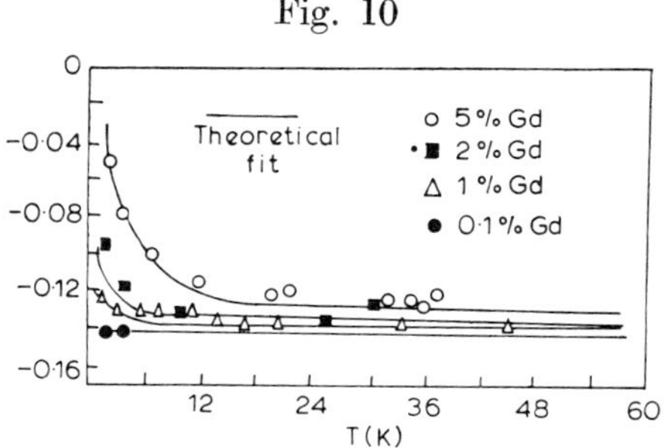

g-value as a function of temperature for the compound $LaNi_5$ with Gd substituted on the La sites (Davidov and Shaltiel, A 38).

The main features of the results are as follows:

1. For Gd concentrations of less than 1%, the g-shift was almost temperature independent but for higher concentrations a shift back to $g = 2$ sets in at temperatures, increasing with increasing concentration.

2. The linewidth at 4 K displayed a strong concentration dependence, increasing from 100 G at the lowest concentration to about 600 G at 3% Gd and then flattening out. The linewidth for each alloy was not plotted explicitly as a function of temperature. There was also a large applied field effect; measurements at X-band (~ 3 KG) giving a linewidth about half that at Q-band (14 kG).

3. The g-shift measurements also showed a pronounced field effect. The g-shift was greater at X-band and the largest differences were observed for high concentrations.

Dynamic effects were again invoked to explain the behaviour in these two systems, although the increase in linewidth with concentration and field was not explained by this hypothesis, rather, it was assumed that local clusters of Gd resulted in line broadening but that exchange narrowing at higher concentration accounted for the subsequent flattening out of the linewidth versus concentration curve.

Recently, Male and Taylor (A 93) have rejected this interpretation of the $La_{1-x}Gd_xNi_5$ data and, by implication, that of the $La_{1-x}Gd_xRu_2$ data as well. They show that for slightly larger concentrations of Gd, the low temperature g-value changed sign to become positive and that for their X-band measurements the dynamic effect expression does not fit their data. They suggest that the temperature, frequency and concentration dependence of both the g-value and linewidth can be explained by invoking the effects of magnetic ordering as the Gd concentration is increased. For an 8% Gd alloy, a good fit may be made to the g-value data using the temperature dependence of the magnetization in $3\,kG$ and assuming a demagnetizing factor ~ 2.8 for the particles of their powder. They show that the g-shift is reduced rather more quickly for the first few per cent of Gd and in this concentration regime the Curie temperature rises more rapidly than at higher concentrations; these effects are attributed to exhaustion of the response of the $LaNi_5$ d-band on increasing the Gd concentration.

The weak ferromagnet $ZrZn_2$ was studied by Davidov *et al.* (A 36), substituting Gd (3%–6%) on the Zr sites. Susceptibility measurements on the compound containing 6% Gd gave $T_c = 20\,K$ and $\theta = 23\,K$. The E.S.R. measurements showed a broad maximum in the linewidth at $40\,K$, decreasing to a minimum at $60\,K$, with a high temperature slope of $3\,G\,.\,K^{-1}$. A negative g-shift was observed with a maximum at $30\,K$ and a shallow minimum at $50\,K$, decreasing asymptotically to $g = 2$ at higher temperatures. Fairly good correspondence was obtained between the shape of the curves of g versus T and ΔH versus T for different concentrations, although for lower concentrations the absolute value of the linewidth was less and the minimum in the g-value more pronounced. The authors claimed that a fairly good fit to these rather curious data could, above $20\,K$, be obtained by consideration of dynamic effects.

Davidov *et al.* (A 37) made measurements on the compound UPd_3 with Th substituted on the U sites in addition to 6% Gd. They claimed that existence of a spin-compensated state of the uranium could explain both susceptibility and their E.S.R. data.

The study of bottlenecking in the dialuminides YAl_2, $LaAl_2$ and $LuAl_2$ with Gd or $Eu(4f^7)$ substituted on the Y, La or Lu sites has been the subject of a considerable number of investigations. Schäfer *et al.* (A 130, A 132) found that Gd in YAl_2 is partially bottlenecked even at low concentrations of Gd. Whilst the slope of the linewidth increases with decreasing Gd concentration, the g-value remained approximately concentration independent at 1.990 ± 0.005, this, presumably, reflecting the linear dependence of Δg on the bottleneck parameter compared to the quadratic dependence of $d\Delta H/dt$ (see § 2). Addition of Th confirmed the presence of a bottleneck in the system, and it was found possible to remove this by a combination of low Gd and high Th concentrations. In the unbottlenecked regime, the g-shift of $\Delta g = 0.07 \pm 0.02$ was roughly twice that expected from the simple Korringa relation. It was suggested that the discrepancy was due to exchange enhancement of the host, although for YAl_2 the differences in $J(q)$ measured by the linewidth and g-shift would perhaps seem a more likely explanation.

Koopmann *et al.* (A 90) examined the effects of bottlenecking of both Gd and $Eu(4f^7)$ in $LaAl_2$ (figs. 11 and 12). At the unbottlenecked limit a g-shift of $+0.05$ was obtained for Eu in $LaAl_2$ and $+0.10$ for Gd in $LaAl_2$. As in the case of YAl_2–Gd (above) these values were about two times the values expected by substituting the linewidth slopes ($26\,\mathrm{G.K^{-1}}$ and $52\,\mathrm{G.K^{-1}}$) in the simple Korringa formula. The authors claimed that the fact that the values of J ($+0.07$ and $+0.13\,\mathrm{eV}$) obtained from Δg were larger than those obtained from the depression of T_c was evidence for the important role of the 5d electrons in the superconductivity as against their relatively unimportant role in the total exchange. This latter analysis once again seems to neglect the $J(q)$ dependence of different measurements.

Fig. 11

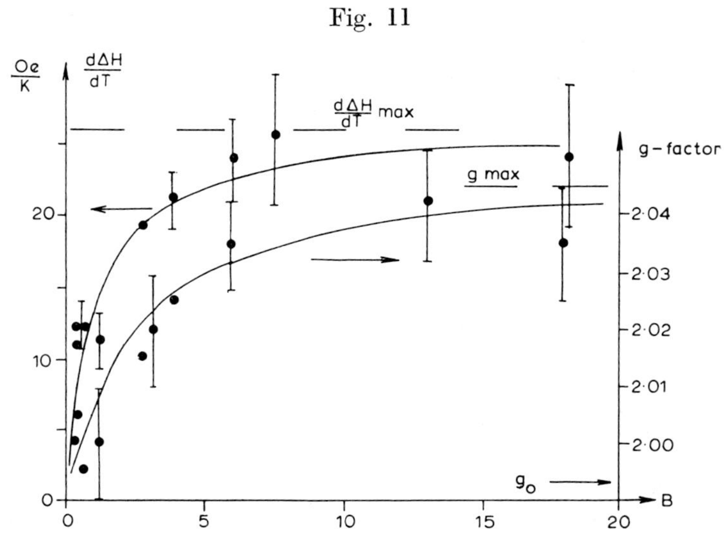

g-factor and slope of linewidth $d\Delta H/dT$ of $\mathrm{Eu}_x\mathrm{La}_{1-x}\mathrm{Al}_2$ as a function of the bottleneck factor $B(=\delta s\mathrm{L}/\delta s\mathrm{f})$ (after Koopmann $et\ al.$, A 90).

Fig. 12

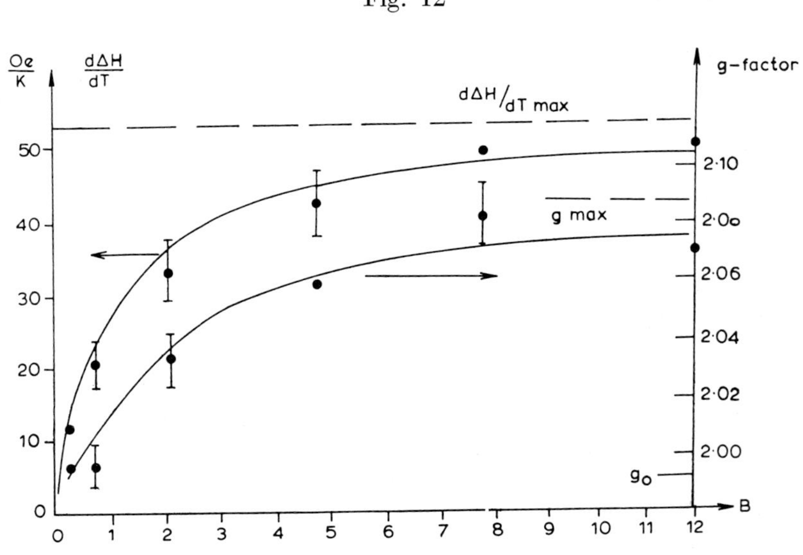

g-factor and slope of linewidth $d\Delta H/dT$ of $\mathrm{Gd}_x\mathrm{La}_{1-x}\mathrm{Al}_2$ as a function of the bottleneck factor B (after Koopmann $et\ al.$, A 90)

Davidov *et al.* (A 56) also investigated the role of the E.S.R. bottleneck in the $La_{1-x}Gd_xAl_2$ system. These authors noticed a large variation in E.S.R. parameters, depending upon the details of sample preparation, but were able to extract clear evidence for bottlenecking despite this. At the dilute limit a value of $g = 2.07$ and a Korringa slope of $60 \, G \, . \, K^{-1}$ were obtained, but addition of Th increased these latter values to 2.11 ± 0.01 and $70 \, G \, . \, K^{-1}$. They looked for, but were unable to provide clear evidence for, the existence of dynamic effects in the system. In this work, Davidov and his co-workers reported a low-field signal which they attributed to superconductivity. By correlating the depression of T_c, the values of Δg and $d\Delta H/dT$ and the upper critical field, they were able to show a proportionality between Δg (and $d\Delta H/dT$) and $1/\tau_{tr}$, where τ_{tr} is the transport collision time for the alloy. Parameters which determine superconducting critical fields and temperature behaviour were thus deduced from the resonance experiments. The suggestion was made in this paper that the small negative shift for $GdAl_2$ compared to the positive shift at the dilute Gd in $LaAl_2$ end of the system could be explained on the basis of a two-band model for bottlenecking where the negative g-shift arising from the d-like character at the Fermi surface could not be bottlenecked and that therefore the change in sign of Δg derived entirely from the opening and closing of the s-electron bottleneck superimposed on a constant negative d-electron background. Using the g-shift figures, the authors obtained values for the exchange interactions J_{f-s} and J_{f-d} of $+0.13 \pm 0.01 \, eV$ and $-0.035 \, eV$ respectively, allowing in the latter case for the effects of exchange enhancement. Bearing in mind the absence of a bottleneck in strongly d-like materials, the two-band model would seem to be a very reasonable approach to this problem. The $LaAl_2$–Gd system with its exceedingly small negative shift at high Gd concentrations would not seem to be the ideal alloy system to apply this concept—it would certainly explain the much larger effects observed in the equiatomic Gd noble metal alloys (see below). As a corollary, application of this model would presumably imply a small amount of s-electron bottlenecking in strongly d-like materials such as Pd and $LaNi_5$, leading to a very small temperature-dependent contribution to Δg. This would be probably masked by interaction effects, however.

Following up the report of Coles *et al.* (A 28) that $LaAl_2 - 0.45\%$ Gd had a positive g-shift ($\Delta g = +0.027 \pm 0.01$), Taylor (A 149) also showed the clear presence of the E.S.R. bottleneck in $LaAl_2$, obtaining a positive g-shift to $+0.063 \pm 0.01$ in the dilute limit and a Korringa slope of $60 \, G \, . \, K^{-1}$. At concentrations above 5% Gd, the effects of magnetic ordering were clearly shown with the linewidth broadening well above the measured Curie temperature (in a 5% alloy, $T_g = 5 \, K$, $T_{\Delta H} = 15 \, K$). It is to be stressed that although these several reports of bottlenecking in the dialuminides would appear repetitive, the various publications appeared within a few months of one another and were the result of a number of independent experimental investigations.

Although dynamic effects were not observed in any of the above measurements, Rettori and his co-workers (A 124) claim to have observed such effects for Gd in $LuAl_2$ (fig. 13). The bottleneck in this system was shown most clearly. Below 2000 p.p.m. Gd, a high temperature ($> 5 \, K$) slope of $73 \, G \, . \, K^{-1}$ was observed but at lower temperatures the linewidth flattens out and increases again for alloys containing more than 1 at. $\%$ Gd. The g-shift is temperature

Fig. 13

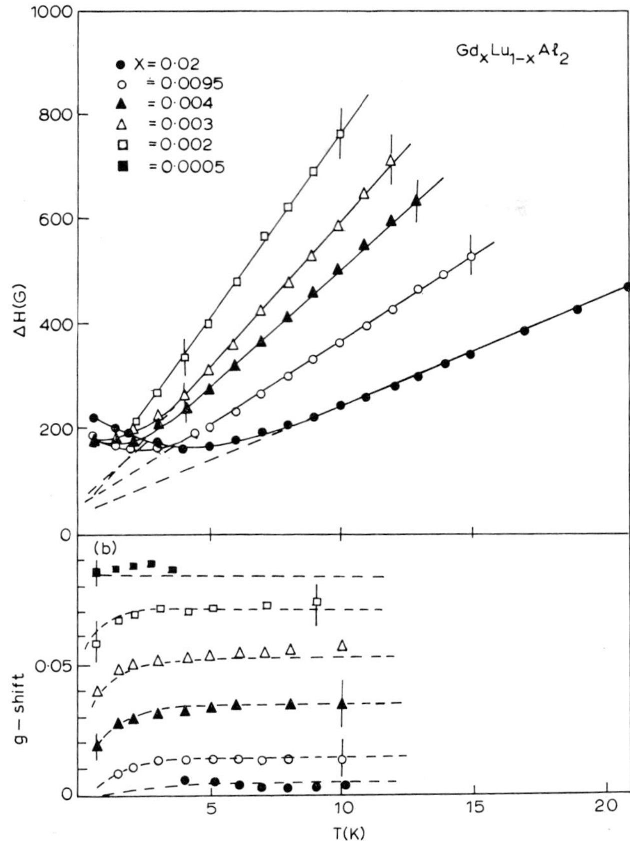

Linewidth and g-shift as a function of temperature for Gd in $LuAl_2$ (after **Rettori** *et al.*, A 124).

independent above 6 K for Gd concentrations in excess of 2000 p.p.m. but below 6 K a decrease in Δg with decreasing temperature was observed. For the lowest concentrations the g-value was found to be concentration and temperature independent with $\Delta g = +0.085$. Addition of Th opened the bottleneck for high Gd concentrations and ordering effects giving rise to an increase in the effective Δg were observed for Gd concentration in excess of 1%. The fact that the linewidth, flattened out for very low concentrations at low temperatures where the g-shift was temperature independent, was attributed to either the effects of further unknown impurities or to unresolved fine structure. Extending previous measurements on YAl_2 down to 0·7 K in a He_3 cryostat and using Gd concentrations of 0·025%, Rettori and his co-workers claim to have identified signs of dynamic effects in this system also. They claim that the non-observation of similar effect in $LaAl_2$–Gd is to be understood in terms of the existence of superconductivity in this system. The q-dependence of the exchange interaction was analysed using the model of Davidov *et al.* (A 48), and values for the spin-flip scattering rates of the conduction electrons caused by the Gd and Th were extracted.

In a subsequent paper, Davidov *et al.* (A 47) added Ce, Th and U impurities to LaAl$_2$–Gd, obtaining $\partial \delta_{\mathrm{SL}}/\partial_c = 7 \pm 3$ for Ce, 5 ± 2 for Th and 20 ± 8 for U. They claim that the larger cross sections for Ce and U compared to Th (and Gd) are due to the greater proximity of the 4f and 5f resonance levels to the Fermi level in the former cases, thus leading to a more rapid spin–orbit relaxation of the conduction electrons.

Bottlenecking in the Y$_{1-x}$Gd$_x$Ag alloy system was found by Weimann *et al.* (A 156, A 157). In the unbottlenecked limit a g-value of 2·067 was obtained and a Korringa slope of 14 G . K^{-1}. A progressive increase in the residual linewidth was also found from about 150 G in the dilute limit to 380 G at 2 at. % Gd. The results for the g-value and linewidth slopes were found to be the same at both X-band and Q-band and no signs of dynamic effects were observed. In the authors' first paper, they point out that the values of J_{eff} obtained from Δg and $d\Delta H/dT$ differ by a factor of 10 and relate this to the q-dependence of J. In the second paper the two-band bottlenecking model of Davidov *et al.* is used to obtain rough values of 0·3 eV and $-0\cdot03$ eV for $J_{\mathrm{s-f}}$ and $J_{\mathrm{f-d}}$. In the same alloy system and in the LaAg and YCu systems, Davidov *et al.* (A 48) applied their partial wave analysis to obtain estimates of $J^{(0)}$ and $J^{(1)}$. They made the assumptions: (1) that exchange enhancement is negligible in these alloys and (2) rather more dubiously that in YCu the effect of the covalent mixing terms $J_{\mathrm{CM}}^{(2)}$ can be entirely neglected with Gd(4f^7) as impurity. With Gd concentrations of from 40 to 2000 p.p.m. Gd and with second impurities (where necessary) to completely break the bottleneck, they obtained the following average values for the three alloy systems. :

LaAg	$\Delta g = 2\cdot070 \pm 0\cdot004$	$d\Delta H/dT = 29 \pm 4$
YCu	$\Delta g = 2\cdot050 \pm 0\cdot004$	$d\Delta H/dT = 18 \pm 4$
YAg	$\Delta g = 2\cdot056 \pm 0\cdot004$	$d\Delta H/dT = 20 \pm 3$

Sperlich *et al.* (A 142) showed that bottlenecking is also present in the compound GdB$_6$ (fig. 14). The high temperature slope of the linewidth (above about 620 K) for the pure compound was estimated to be $0\cdot6 \pm 0\cdot1$ G . K^{-1} and this increased to 2·4 G . K^{-1} for (La$_{80}$Gd$_{20}$)B$_6$ and 9·1 G . K^{-1} for (La$_{90}$Gd$_{10}$)B$_6$. By also measuring the exchange narrowed linewidth and calculating the second moment for a line-broadening mechanism such as dipole–dipole interactions, an estimate was made for the exchange field. It was estimated that this fell from 69 ± 5 kG in GdB$_6$ to 21 ± 5 kG for (La$_{90}$Gd$_{10}$)B$_6$. The value for H_{E} in GdB$_6$ was checked by using it to estimate the exchange parameter between Gd ions at the distance r_0 and through the RKKY formalism to compare calculated and measured values of θ.

Eu and Gd were diluted into the compounds SmB$_6$, SrB$_6$ and LaB$_6$ by Kojima *et al.* (A 89) obtaining the g-values:

	Gd	Eu(4f^7)
SrB$_6$	$1\cdot998 \pm 0\cdot003$	$1\cdot989 \pm 0\cdot003$
LaB$_6$	$1\cdot995$	—
SmB$_6$	$1\cdot925 \pm 0\cdot005$	$1\cdot937 \pm 0\cdot005$

The noticeably large negative g-shift for Gd and Eu in SmB$_6$, it was suggested, may be as a result of a strong interaction between the 4f^7 bound state and the Sm 5d electrons.

Fig. 14

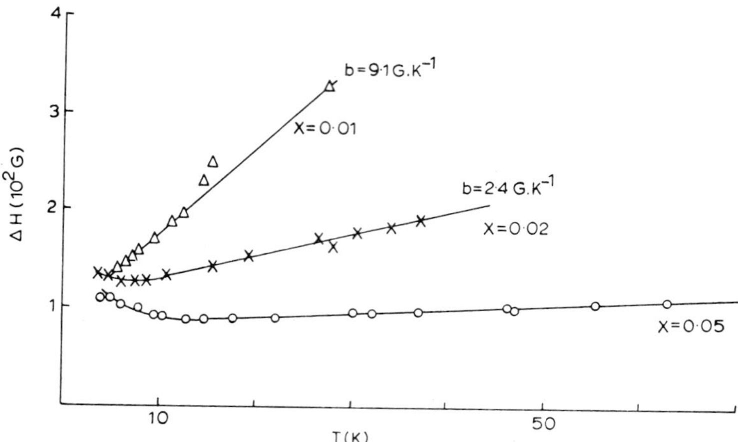

E.S.R. on ·polycrystalline $Gd_xLa_{1-x}B_6$—temperature dependence of the linewidth ΔH at 35 GHz for $x=0\cdot0$, $0\cdot02$ and $0\cdot05$ (Sperlich *et al.*, A 142).

Fine structure of s-state ions in intermetallic compounds has now been resolved in two cases. The first observation by Davidov *et al.* (A 55) of Gd in concentrations of from 250 to 3000 p.p.m. in a single crystal of LaSb yielded a value of the fourth-order crystal-field parameter b_4 (See above, § 2) of 30 ± 2 G. A g-value of $1\cdot987 \pm 0\cdot005$ with a thermal broadening of the $\frac{1}{2}\leftrightarrow-\frac{1}{2}$ fine-structure line of $1 \pm 0\cdot5$ G . K⁻¹ was obtained for the 500 p.p.m. alloy. Unlike the case of Pd–Gd, no significant narrowing or broadening effects were found as a function of temperature, but appreciable narrowing occurred as a function of increasing Gd concentration. In the 3000 p.p.m. alloy, indeed, almost complete exchange narrowing had been achieved, the authors estimating an exchange field of ~ 200 G for this nominal concentration. Elschner *et al.* (A 72), in a short communication, announced observation of well-resolved fine structure of $Eu(4f^7)$ in single crystals of $BaAl_4$ with Eu concentrations of 1% and 0·5%.

The E.S.R. behaviour of Gd in $CeRu_2$ at 4·2 K, 77 K and 25 °C was investigated by Sun and Schnitzke (A 142). At 4·2 K the g-value was found to be slightly less than the Gd free ion value but Δg was found to change sign on the addition of about 9% Gd. This latter effect was attributed to the effects of magnetic ordering. The g-value at 77 K was found to change sign for a 60% addition of Gd where the linewidth was also narrowest. Most of the studies of E.S.R. in this compound and other $-Ru_2$ alloys have been concerned with resonance in these alloys as type II superconductors. The first observation† of E.S.R. in a superconductor was reported by Al'tshuler *et al.*, on a sample of 0·25 at. % Gd in La_3In. A g-value of $2\cdot005 \pm 0\cdot005$ was measured but no explicit transition was observed at the superconducting–normal state boundary. However, in this respect, measurements in the $-Ru_2$ compounds were more successful, Rettori *et al.* (A 123) found that the g-value in $LaRu_2$ changed from $1\cdot82 \pm 0\cdot005$ in the normal state to $1\cdot84 \pm 0\cdot005$ in the superconducting state with a change in

---

† See comment on foot of p. 701.

lineshape ratio from 2 to 1·2 below the transition temperature. Even more convincingly, the slope of the linewidth increased from $23 \pm 4\,\mathrm{G}\,.\,\mathrm{K}^{-1}$ in the normal region to $50 \pm 10\,\mathrm{G}\,.\,\mathrm{K}^{-1}$ in the superconducting state. This is shown in fig. 15. The change in g-shift at the transition temperature was found to decrease with increasing Gd concentration.

Fig. 15

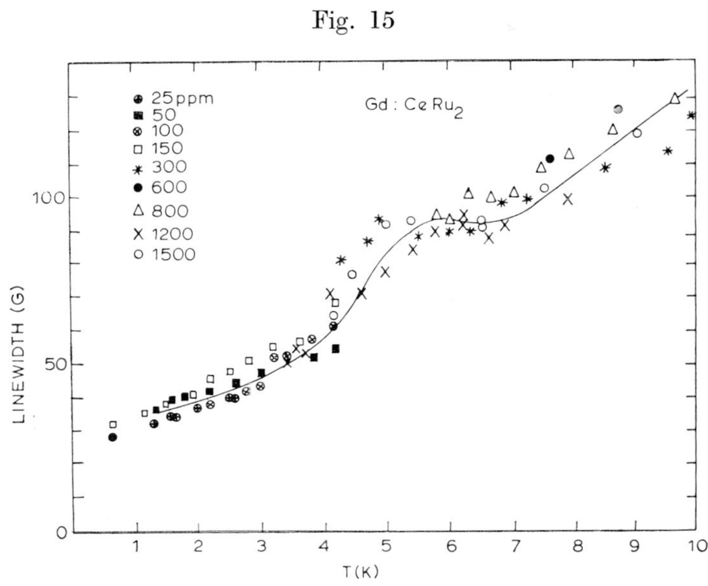

E.P.R. linewidth of Gd in CeRu$_2$ as a function of temperature using expanded ordinate. Data for high Gd concentrations at low temperatures are omitted (after Rettori *et al.*, A 46).

From the E.S.R. measurements the authors were able to extract values of $1/\tau_{\mathrm{so}}$, the spin-orbit scattering rate, and of $1/\tau_{\mathrm{ex}}$, the conduction electron exchange scattering rate. The same parameters were also obtained from direct measurements of the re-entrant critical field and the reduction in T_{c} on alloying with Gd. Despite the rather gross assumption of a free-electron Fermi surface, order of magnitude agreement was obtained. The authors pointed out that the sign of the g-shift in the superconducting state supports other evidence that LaRu$_2$ is a d-band superconductor. Although LaRu$_2$ is unbottlenecked, the possibility was put forward that E.S.R. in bottlenecked systems such as LaAl$_2$ −Gd could be observed in the superconducting state and that in this situation application of the theory of Maki (1969), which suggests that $1/\tau_{\mathrm{so}}$ is reduced whilst $1/\tau_{\mathrm{ex}}$ is increased at the transition, would imply the possibility that a change from unbottlenecked ($1/\tau_{\mathrm{so}} > 1/\tau_{\mathrm{ex}}$) to bottlenecked ($1/\tau_{\mathrm{ex}} > 1/\tau_{\mathrm{so}}$) could be observed in principle by cooling through the transition temperature.

Working independently Engel *et al.* (A 73) also observed E.S.R. in the normal and superconducting regimes for Gd in CeRu$_2$ and LaRu$_2$. For Gd in CeRu$_2$ with concentrations of 80, 300 and 500 p.p.m., a thermal broadening in the normal region of $4\,\mathrm{G}\,.\,\mathrm{K}^{-1}$ was observed, compared with a value of $43\,\mathrm{G}\,.\,\mathrm{K}^{-1}$ just below the transition temperature ($\sim 6\,\mathrm{K}$ in the applied field), but the linewidth flattened out as a function of temperature below, about 3 K. The

g-value was also found to shift from $1\cdot965 \pm 0\cdot005$ in the normal state to $1\cdot944 \pm 0\cdot005$ below T_c. These results were compared with measurements on Gd in $LaRu_2$ in which $g_n = 1\cdot830 \pm 0\cdot005$ and $g_s = 1\cdot845 \pm 0\cdot005$ (where the subscripts refer to the normal and superconducting values), these values being in good agreement with the results of Rettori $et~al.$ The change in sign of $g_s - g_n$ between the two systems is extremely interesting and was thought to suggest a change in the proximity of the d-electrons to the Fermi surface.

The form of the g-shift and thermal broadening of 250 p.p.m. of Gd in $Ce_xLa_{1-x}Ru_2$ as a function of X in the normal region was analysed by Baberschke $et~al.$ (A 8). For $0 \le X \le 50$, no change in Δg and $d\Delta H/dT$ from the $LaRu_2$ value was observed (although the value for $d\Delta H/dT$ of $4\,G \cdot K^{-1}$ quoted in the previous paper was here amended to $\sim 10\,G \cdot K^{-1}$) but for $60 \le X \le 100$, $d\Delta H/dT$ increased monotonically to $24\,G \cdot K^{-1}$ and g dropped to $1\cdot85$. Making the assumptions implicit in the use of the simple Korringa relation ($J^2(o) = \langle J^2(q) \rangle$, no enhancement and a single conduction electron band) a value of $\Delta g^2/d\Delta H/dT$ is obtained, which is in good agreement with experiment for $0 \le X \le 50$; for higher concentrations of La, however, poor agreement is obtained and the authors speculate that the discrepancy may be due to an additional f-band at the Fermi surface as the calculations of Switendick (1973) for $LaAl_2$ would suggest. However, as Baberschke $et~al.$, themselves point out, the assumptions made above may vary in their validity across the alloy system.

Superconductivity and E.S.R. measurements of Gd in Th–, Ce– and $LaRu_2$ were described by Davidov $et~al.$ (A 46). At low concentrations (< 1000 p.p.m.) normal state g-shifts of $-0\cdot035$, $-0\cdot05$ and $-0\cdot175$ respectively were obtained and thermal broadenings of 7 ± 2, 12 ± 3 and $23 \pm 5\,G \cdot K^{-1}$. Measurements of T_c as a function of Gd concentration showed a slight rise initially for $Gd_xTh_{1-x}Ru_2$ and then remained roughly constant (at about $3\cdot2\,K$) up to a Gd concentration of 6%. This behaviour was explained by assuming that the Gd in this system is acting chiefly as a potential scattering impurity with a much reduced scattering by virtue of its wave-vector dependence. By adding La to $ThRu_2$ and subtracting these data from the data for the Gd alloy, an upper limit on the depression of T_c due to the Gd was obtained at $0\cdot2\,K$ at. $\%$ Gd^{-1}. A similar estimate was made for Gd in $CeRu_2$. Assuming that the same electrons responsible for the superconductivity are also responsible for the thermal broadening (both of which are related to $\langle |J(q)|^2 \rangle$) and choosing $LaAl_2$ as a standard, an estimate of $(\Delta T_c/\Delta n)_{ex}$ for $LaRu_2$, $ThRu_2$ and $CeRu_2$ containing Gd was obtained with measured values. In a further paper (A 46), a correlation between the E.S.R. g-shift and the re-entrant critical-field behaviour associated with Pauli pair breaking for Gd in $ThRu_2$ was suggested. Measurements of H_{c2} versus temperature for different Gd concentrations showed that H_{c2} decreases with decreasing temperature below about 1 or $2\,K$. From the Abrikosov–Gor'kov theory (1962), extended by Fulde and Maki (1966) and using the measured g-shift (see above) to give a value of $J(o) = -0\cdot017\,eV$, a value of $1/\tau_{so} = 8 \pm 4 \times 10^{13}\,sec^{-1}$ was obtained for the spin–orbit scattering rate for 6% Gd in $ThRu_2$.

A more detailed discussion of the results of E.S.R. in the superconductors $LaRu_2$, $CeRu_2$ and $ThRu_2$ is given by Davidov $et~al.$ (A 51) including the results obtained on extending their measurements into the pumped He^3 temperature range. In $ThRu_2$, the values of $(d\Delta H/dT)_n$ and $(d\Delta H/dT)_s$ appeared to be the same well away from the transition region, as in the case of the other two com-pounds. The change in g-value for Gd in $ThRu_2$ was smaller than the error limits of the measurement, but was thought probably to be of the same sign as in $CeRu_2$. An argument due to Maki (unpublished) was used to try to account for the drop in the A/B ratio on entering the superconducting state and various hypotheses were advanced to attempt to explain the reason for the puzzling change of sign of g_s–g_N for $LaRu_2(<0)$ and $CeRu_2(>0)$. However, this problem seems to have been overcome by Baberschke *et al.* (A 7). They have shown that the sign of the change in g-value at the superconducting transition for Gd in $CeRu_2$ is different, depending upon the applied field. In the normal state a tem-perature and field-independent g-value of 1·95 was obtained, but on going below the transition at X-band (3·4 kG) the g-value falls to 1·93, whilst for Q-band (12·7 kG) it increases to 1·96 (fig. 16). This change in g-factor at the transition, it was concluded, must not only be produced by a decrease in the spin suscepti-bility but also by a change in the internal field for the vortex state. At X-band, the internal field distribution is wide and thus the saddle-point field, H_s, which the majority of the lattice points experience in the vortex state, is reduced as compared to the applied field. As the temperature is lowered, Baberschke *et al.*, argue that the field distribution widens, decreasing the g-factor further and outweighing the decrease in spin susceptibility. At the higher fields corre-sponding to Q-band measurements, however, the internal field distribution is narrower and the decrease in spin susceptibility is the dominant contribution. The fact that in $La_{1-x}Gd_xRu_2$ an increase in g-factor is observed for low fields would seem to imply that there is a larger decrease in spin susceptibility in this compound than in $CeRu_2$. These arguments seem to dispense with the need for the complex two-band models suggested by Davidov and his co-workers.

Fig. 16

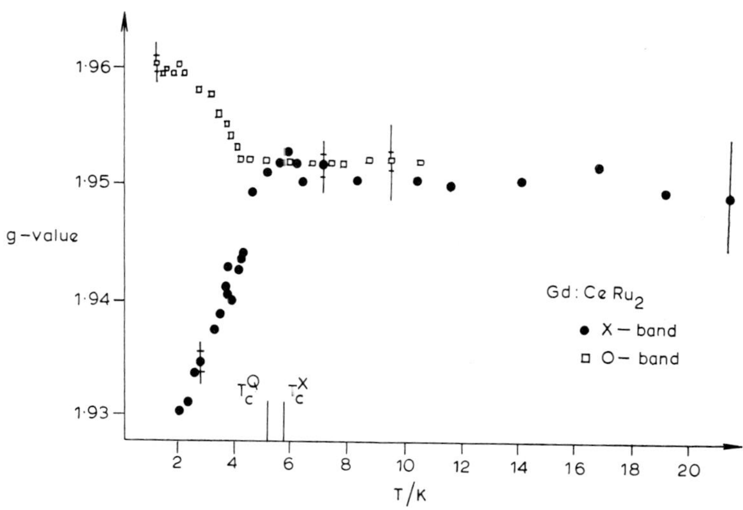

g-value of Gd : $CeRu_2$. The g-value is independent from the Gd concentration. T_c is a function of the applied field H_0. Big error bars include all systematic errors. The small ones are relative error bars (Baberschke *et al.*, A 7).

4.4. *Rare earths in Pd and Pd binary alloys*

The first observation of E.S.R. in the much studied Pd–Gd system was made by Peter *et al.* (A 115).   They found $g = 1.887 \pm 0.007$ for 3% Gd in Pd at 20 K and plotted the linewidth against temperature at different frequencies.   At high temperatures a linear relation was observed with a slope of about $5\,\mathrm{G}\,.\,\mathrm{K}^{-1}$ for 1% and 3% Gd at all frequencies.   At 10 K, 12 K and 20 K they found broad minima for the 1% alloy at 9 GHz, the 3% at 12 GHz and the 3% at 50 GHz respectively, with strong broadening in the higher concentration alloys, from about 600 G at the minimum to 1800 G at 2 K.

On dissolving Gd in a number of hosts from pure Rh ($g = 1.89 \pm 0.007$) through Rh–Pd and Pd–Ag alloys to pure Ag ($g = 1.995 \pm 0.007$), i.e. effectively varying the atomic number of the host from 45·0 (pure Rh) to 47·0 (pure Ag), they found a strong minimum in the g-value measured at 20 K and 78 K, and a peak in the linewidth at a host atomic number slightly on the Rh side of pure Pd. This minimum corresponded to peaks in the electronic specific heat and susceptibility obtained by Budworth *et al.* (1960).   In a further paper, Peter *et al.* (A 114) discussed the effect of adding other rare earths to the Pd–Gd host. They found that (fig. 17) the addition of the light rare-earth elements increased the g-shift, whilst addition of heavy rare earths decreased it.   They also showed, from susceptibility measurements of Tb in Pd and Pr in Pd, that the g-shift was proportional to the susceptibility of the added rare earths.   The measurements suggested that the range of the positive polarization in Pd was far greater than the range of the first change in sign of the RKKY range function in a simple unenhanced host (see § 2).

Fig. 17

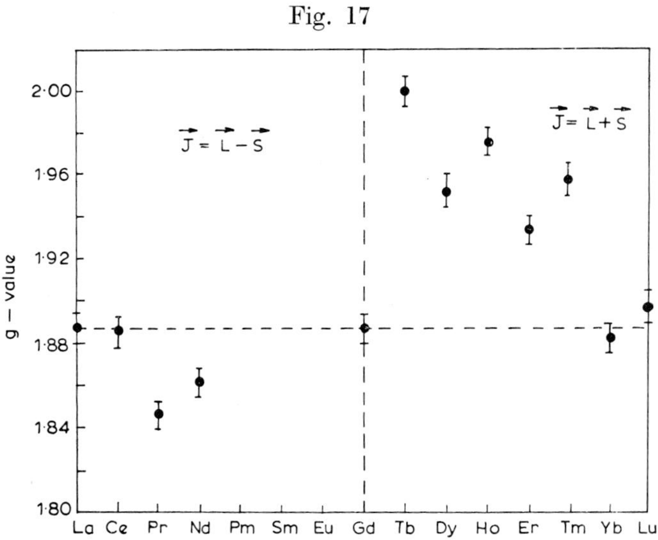

g-values of Gd in the alloys 96% Pd, 2% Gd, 2% rare earths at 20 K (Peter *et al.*, A 114).

Shaltiel *et al.* (A 140) extended their previous work on the system by adding small concentrations of Tb, Sm, Pr and Ce to a Pd 3% Gd alloy and measuring the g-shift as a function of temperature.   Below about 20 K the g-value of the alloy containing Tb shifted to 2·10 whilst that of the Pr alloy dropped to 1·75.

The Sm and Ce-doped alloys showed anomalous behaviour; the Sm behaved as though it were a heavy rare earth, giving a g-value of 2·00 at 2 K, whilst no additional shift was observed in the case of Ce. The g-value of Gd in Pd was found to be approximately concentration independent over the range 1% to 3% Gd, but addition of Tb and Nd to an alloy containing 2% Gd gave a g-value which increased with increasing Tb content and decreased on the addition of Nd. When Ni was added to a Pd 3% Gd alloy, the g-shift and linewidth were found to increase linearly with the Ni concentration. Up to 0·3 at. % of Fe and Co were also added and in the case of Co impurities the g-shift at 20 K increased from the value observed for Gd in pure Pd but became smaller than the Pd–Gd g-shift as the temperature was lowered. The first of these effects was attributed to an increase in the Pauli susceptibility (as in the Ni case), whilst the second effect was held to suggest a negative sign for the exchange interaction between the Co ion and the conduction electrons. No resonance from the Co was ever observed, whereas the measurements on alloys containing Fe were hampered by a strong Fe ferromagnetic resonance signal.

In the same paper, E.S.R. was also reported for alloys of 2% Gd diluted into a series of Pd–U hosts, containing up to 20% U. On the addition of U to the pure Pd–Gd alloy, the g-value increased towards $g = 2$, reaching this value at 9% U, at which concentration there also exists a minimum in the Pd–U susceptibility. When the Pd–U g-shift data were plotted on the same concentration scale as Pd–Ag, the two sets of values were found to correspond. The linewidth in the Pd–U–Gd system was found to be roughly independent of concentration up to about 9%, but was then found to increase rapidly from 400 G to 1200 G at 15%.

In a further short note, the band-filling effects of a third element were shown up very clearly when Shaltiel (A 136) hydrogenated his Pd–Gd alloys. The g-value returned to $g = 2$ as expected and the addition of other rare earths led to no further shift.

Fig. 18

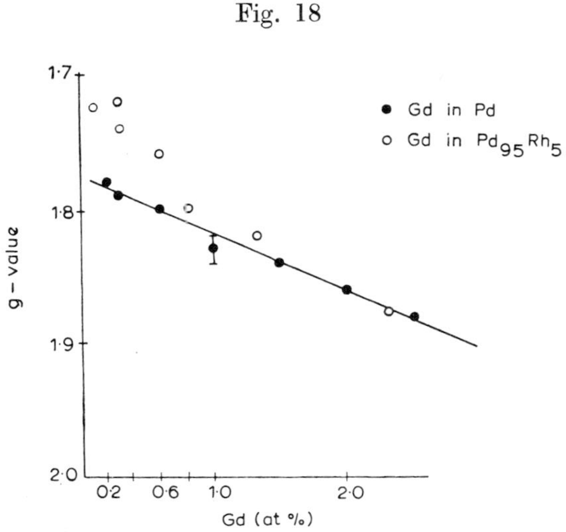

E.P.R. g-value versus Gd concentration for Gd in Pd and Pd$_{95}$Rh$_5$ (Cottet and Peter, A 32).

Fig. 19

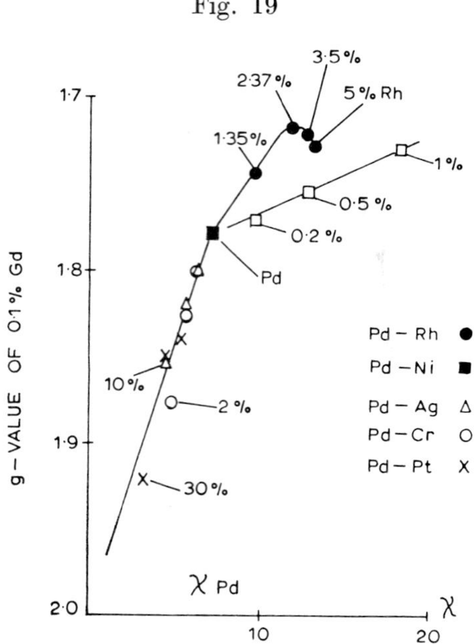

E.P.R. g-value for 0·1% Gd in alloys of Pd with Ni, Rh, Ag, Cr, Pt at 4·2 K (Cottet and Peter, A 32).

Cottet and Peter (A 32) (figs. 18, 19) worked extensively on the Pd–Gd and binary Pd–Gd systems. They found that for 0·1% Gd, the g-value is temperature independent at $g = 1·78$ and that this increases linearly with concentration to 1·88 for 2·5 at. % Gd in Pd. Cottet (A 35) plotted a family of curves for both ΔH versus T and g versus T for concentrations of 0·1, 0·4, 0·6 and 1 at. % Gd in Pd. In the high temperature regime (> 30 K) the linewidth increases linearly with a slope of about 6 G K^{-1} and each curve shows a fairly strong minimum between 10 K (0·1%) and about 30 K (1·0%). Above about 0·4%, the g-value shows a slight curvature towards higher g-values at low temperatures. Addition of Ni or Rh to the Pd host gave an increased g-shift to $g \sim 1·73$ for 1% Ni or 2·5% Rh. It was found that for Pd alloys doped with Cr, Pt and Ag, where the susceptibility of the host is lower than in pure Pd, the g-value was proportional to the host susceptibility at 4·2 K. The same was true for Ni-doped alloys where the suceptibility is higher than for pure Pd, but in this case a new constant of proportionality was found. In the case of Rh, however, this proportionality was no longer observed. When the g-shift was plotted against the high temperature (250 K) susceptibility, simple proportionality was observed for all the alloys. Cottet and Peter suggested that both Ni in Pd and Rh in Pd are partially localized and that the temperature-independent part of these solutes in Pd corresponds to the delocalized part of the susceptibility which largely controls the g-shift. The localized part influences the linewidth and here it was found that whilst addition of Ni and Rh broadened the line considerably, addition of Pt and Ag had little effect.

Praddaude and Gärtner (A 121) measured the g-values of 0·2–2% Gd in Pd using specimens in the form of thin plates. Their lowest observed g-value was 1·826 and they suggested that the lower values of Cottet and Peter were obtained on the basis of powder work where demagnetization corrections are more difficult to apply. However, inclusion of a demagnetizing factor to the results of Cottet and Peter would further increase the discrepancy.

The measurements of Coles *et al.* (A 28) and Taylor and Coles (A 150) support strongly the measured values of Cottet. In this work the g-values and linewidths of 0·02 at. % to 6 at. % Gd in Pd are plotted as a function of temperature (except in the two most dilute alloys where values at 1·7 K and 4·2 K only are given) and the role of magnetic ordering on the g-value and linewidth made clear. In the extreme dilute limit a g-value of $1·77 \pm 0·01$ was obtained and this was found to increase to $1·95 \pm 0·01$ at 6 at. %. These results and the further addition of Pt to one of the alloys showed that the system is not bottlenecked at any temperature or concentration. The variation of g-value with increasing Gd content was interpreted as evidence of the filling of the Pd 4d holes by the solute valence electrons and support for this view was taken from the results of Gd in Pd–Ag and Pd–Lu obtained by Cottet (A 35), which showed the same general form of concentration dependence. It was pointed out in this work that there exists a considerable problem in reconciling the values of $\Delta g/g = -11\%$ found in E.S.R. measurements, the reductions from the free Gd ionic moment of the paramagnetic and saturation moments as measured by Shaltiel *et al.* (-22%) (A 140), Schaller *et al.* (-18%) (1972) and Crangle (-20%) (1964) and those obtained by Guertin *et al.* (1971, 1973) where a saturation moment per Gd atom equal to the free Gd^{3+} ion value was obtained. These authors reported an apparent frequency dependence of the residual E.S.R. linewidth in these alloys.

Fully resolved fine structure of Gd in single crystals containing 1300, 650 and 200 p.p.m. Gd was reported by Devine *et al.* (A 65), leading to a value of b_4 of 25 ± 2 G (see figs. 20–22). The resonance line was again reported to be much broader at 35 GHz and it was suggested that this was due to inhomogeneous broadening, resulting from local distortions in the cubic field symmetry; g-values of 1·813, 1·792 and 1·806 all $\pm 0·005$ were reported in ascending order of concentration and a concentration independent slope of the linewidth of $9·3 \pm 0·5$ G . K^{-1} was found. Both the g-values and linewidths showed a small upturn at low temperatures due to the onset of ordering. The width of the fine-structure lines was found to be narrower than that predicted by a simple second-moment calculation and Devine *et al.*, believed that this effect, as in Au–Gd (A 24), arises from the exchange narrowing due to ion–ion exchange. Good agreement between theory and experiment, however, was only obtained when a temperature-dependent ion–ion exchange coupling term, together with an added central resonance, was assumed. Further study of this problem was carried out by Moret *et al.* (A 98) on single crystals containing 300 and 500 p.p.m. of Gd grown by the recrystallization technique. Fine structure was again observed, as reported in the previous study, using crystals grown by zone refining. The results were fitted to the theory of Barnes (1974) and Zimmerman *et al.* (1972) and a value of the cubic crystalline field splitting parameter $b_4 = 29·6$ G was obtained, in good agreement with Devine *et al.* (A 65). It was found that the anomalous line at the $M_s = \frac{1}{2} \leftrightarrow M_s = -\frac{1}{2}$ transition was weaker than in the previous work but still clearly present. These authors suggested

Fig. 20

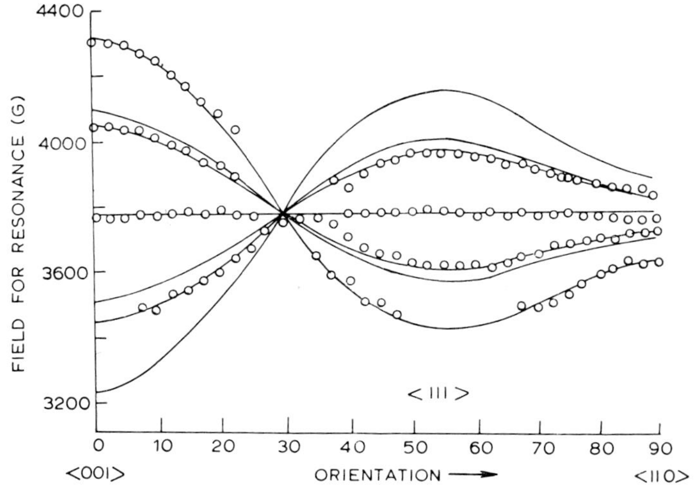

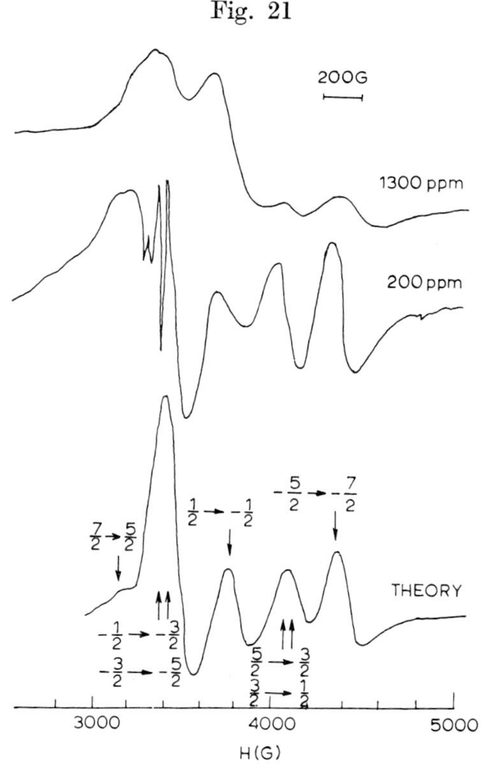

The magnetic field for resonance of the observed lines in the fine structure spectrum of a 200 p.p.m. sample of Gd in Pd. The theoretical results (to first order in b_4) are shown by the solid lines (after Devine *et al.*, A 65).

Fig. 21

The resonance spectra in the (001) direction for 1300 and 200 p.p.m. samples of Gd in Pd, together with the theoretical spectrum with positive b_4 (Devine *et al.*, A 65).

Table 2. Summary of E.S.R. data on the Pd–Gd system

Reference	Conc. (at. %)	Form of sample	Frequency (GHz)	g-value	Temp. (K)	$\dfrac{d\Delta H}{dT}\,\mathrm{GK^{-1}}$	ΔH residual
Peter *et al.* (A 114, A 115)	3	Powders	50	$1\cdot887 \pm 0\cdot007$	20	4	650
	1		50	—	—	—	—
	3		12	—	—	$5\cdot5$	520
	1		9	—	—	—	410
Shaltiel *et al.* (A 140)	1	—	9	$1\cdot886$	20	—	—
	2	—	35	$1\cdot86 \pm 0\cdot03$		—	—
Harris *et al.* (A 81)	$4\cdot4$	Powders and discs	35	$1\cdot86 \pm 0\cdot03$		—	—
	$5\cdot1$			$1\cdot86 \pm 0\cdot03$	60	—	—
	$6\cdot2$			$1\cdot86 \pm 0\cdot03$	—	—	
Cottet and Peter (A 32, A 35)	$0\cdot1$	Powders	35	$1\cdot778$	5	—	—
	$0\cdot4$			$1\cdot790$		—	—
	$0\cdot6$			$1\cdot801$		—	—
	$1\cdot0$			$1\cdot825$		—	—
	$1\cdot5$			$1\cdot84$		—	—
	$2\cdot0$			$1\cdot86$		—	—
	$2\cdot5$			$1\cdot88$		—	—
	$3\cdot0$			$1\cdot892$		—	—

Table 2. (*Continued*)

Praddaude and Gartner (A 121)	0·3	Plates	35	1·83(6)	4·2	40	—
	0·7			1·850		—	—
	1·4			1·86(7)		—	—
	1·9			1·88(7)		—	—
Devine *et al.* (A 65)	0·02	Single crystals	9	1·813	1·47	9·3 ± 0·5	—
	0·065			1·792		9·3 ± 0·5	—
	0·13			1·806		9·3 ± 0·5	—
Taylor and Coles (A 28, A 150)	0·02	Powders	9	1·77 ± 0·02	4·2	—	—
	0·1			1·775 ± 0·010	4·2	—	120 ± 10
	0·2			1·795 ± 0·005	4·2	—	190 ± 10
	0·4			1·815 ± 0·010	4	12 ± 2	200 ± 20
	1·0			1·833 ± 0·010	10	10 ± 1·5	430 ± 20
	4·0			1·93 ± 0·01	10	9 ± 1	530 ± 20
	6·0			1·95 ± 0·01	10	7·5 ± 1	600 ± 20
Moret *et al.* (A 98)	0·03	Single crystals	9	1·788 ± 0·007	1·3	9·3 ± 0·4	—
	0·05			1·788 ± 0·007	1·3	—	—
	0·13			1·813 ± 0·007	1·3	—	—
				1·785 ± 0·007			

Fig. 22

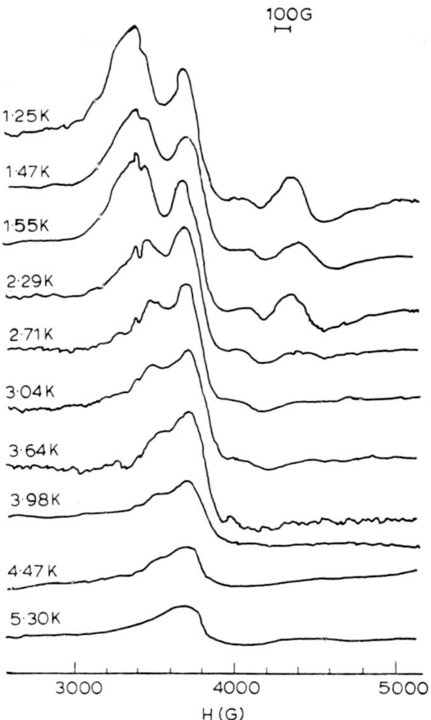

The spectra for the 1300 p.p.m. Gd in Pd sample in the (001) direction as a function of temperature. (Devine *et al.* A65)

that the extra line originated from large spin–spin interactions between nearby Gd sites in the Pd matrix which narrow the fine-structure spectrum into a single line at the centre of gravity. They suggested that the range of this coupling may be equivalent to a sphere containing ~ 200 Pd lattice sites and corresponding approximately to a 0·5% antiferromagnetic conduction electron polarization.

The g-values obtained in this work are, in general, in good agreement with previous results. The values for the $\pm\frac{1}{2}$ line was $1\cdot788 \pm 0\cdot007$ for the 300 p.p.m. and 500 p.p.m. samples and $1\cdot813 \pm 0\cdot007$ for the 1300 p.p.m. sample, whilst the centre of gravity of the field for resonance of the fine-structure lines corresponded to g-values of $1\cdot788 \pm 0\cdot007$, $1\cdot788 \pm 0\cdot007$ and $1\cdot785 \pm 0\cdot007$. This discrepancy between the centre of gravity g-value and the $\frac{1}{2}\leftrightarrow-\frac{1}{2}$ g-value in the most concentrated alloy was attributed to concentration clustering. An attempted partial wave analysis of the exchange parameters obtained from the g-shift and thermal line broadening ($9\cdot3 \pm 0\cdot4\,\mathrm{G.K^{-1}}$ in the 300 p.p.m. sample) showed that such an analysis cannot be applied in the case of an alloy system with such a highly anisotropic Fermi surface.

The results of the studies of the Pd–Gd system reported in this section are summarized in table 2.

4.5. *Gd and Eu(4f⁷) in Y, Lu, Sc, La or Yb*

Popplewell and Tebble (A 118) measured the g-value of Gd–Y alloys containing from 3·7 at. % Gd right across the alloy system in eleven concentration steps to pure Gd. The alloys were found to be concentration dependent, ranging from $1·96 \pm 0·03$ at pure Gd to $2·13 \pm 0·05$ for the 3·7 at. % alloy, passing through the free-ion value at concentrations around 30 at. % Gd. They observed that cooling through the Curie point led to line broadening—the width becoming indeterminate about 100 K below T_c and the line being shifted to lower effective fields. In the alloys with concentrations between 25 at. % and 60 at. % Gd, where in addition to ferromagnetic order there was reported to exist a regime of antiferromagnetic order, two lines were observed. One of these lines broadened and disappeared above T_N, whilst the other appeared to be a ferromagnetic resonance. Popplewell and Tebble suggested that these double lines arose from ferromagnetic clusters of Gd. For a 64% alloy at high temperatures a slope of about 15 G . K⁻¹ was observed in the linewidth and a minimum was found about 40 K above T_c.

In further papers (A 81, A 82) the same authors alloyed Gd to Lu, Sc and La as well as Y. In the Lu–Gd system the g-shift in the paramagnetic regime becomes positive at 80% Gd and for the lowest concentration measured (8·3 at. %) is equal to $2·015 \pm 0·01$. The linewidth again exhibited a minimum some 40 K above the paramagnetic Curie point (220 K) for an alloy containing 70% Gd and continued to broaden below T_c. The large positive g-shift (to 0·13) quoted in the previous study for Gd in Y at the lower concentration end was revised in this work to a figure closer to the free-ion value. The g-value for an alloy containing 8% Gd was given as 1·998. The g-values for 'dilute' Gd in La and Sc were both quoted as being equal to the Gd free-ion value.

Cottet (A 35) examined the behaviour of alloys of Gd in Y, La and Sc, for concentrations less than 2 at. % Gd and quoted the following g-values at 4·2 K :

$$\begin{array}{ll} \text{Y–Gd} & 2·04 \pm 0·02 \\ \text{La–Gd} & 2·03 \pm 0·02 \\ \text{Sc–Gd} & 2·00 \pm 0·02 \end{array}$$

A plot for $d\Delta H/dT$ against concentration in the paramagnetic regime showed an increase in $d\Delta H/dT$ in all three systems but was particularly large in Y–Gd (from 35 G . K⁻¹ at 2% to 100 G . K⁻¹ at 0·2%). Plots of ΔH against temperature for 4·2% and 3·5% Gd in Y showed a broad minimum at 30 K and 20 K respectively, with strong line broadening at low temperatures. The temperatures of the minima in these alloys correspond quite well to the ordering temperature. Cottet concluded from the absence of an appreciable g-shift and from the behaviour of $d\Delta H/dT$ as a function of concentration that the Y–Gd system was well bottlenecked.

The first E.S.R. measurements using single crystals of Gd in this class of host were carried out by Salamon (A 127). In an alloy containing 500 p.p.m. of Gd, he obtained a g-value which varied from 1·988 to 2·07 as a function of orientation with the applied field. The linewidth was found to pass through a minimum at an angle of about 60° between the field and the c-axis. Salamon attempted to explain his results in terms of an anisotropic g-value, but Tao *et al.* (A 147) later showed that the form of the results could be best interpreted in terms of unresolved fine structure.

Schäfer *et al.* (A 129, A 131) dissolved up to 10% Eu in Yb. A signal was observed from 1·6 K to 77 K and measurements were made at both X-band and Q-band. A value of $g = 1·970 \pm 0·005$ was obtained at all temperatures and the alloys all showed the same slope of the linewidth (24 G . K^{-1}) above about 30 K; below this temperature the linewidths passed through a broad minimum upon magnetic ordering. The values of J_{eff} calculated from the g-shift and thermal broadening were in good agreement. In a further study of these alloys Schmidt and his co-workers (A 133, A 134) provided one of the clearest illustrations of dynamic effects in the E.S.R. bottleneck. A strong temperature dependence of the g-value was observed for 0·5% Eu in an alloy of Yb doped with 2% Ca (fig. 23). At 35 GHz (Q-band) the g-value shifted from 2·07 at 5 K to $g = 2·12$ at 20 K and was well-fitted by the full bottleneck formula including the terms responsible for the dynamic effects. A fairly good fit was also obtained for data on the same alloy at 9·6 GHz (X-band), where, as predicted, a larger g-shift as a function of temperature was observed. The dynamical effects in these alloys may, of course, be clearly distinguished from the effects of magnetic ordering which would lead to an increase in Δg as the temperature is lowered.

Fig. 23

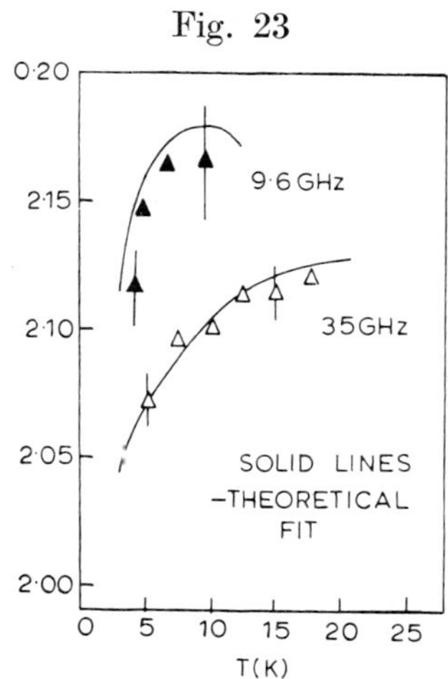

The g-value of 0·5% Eu in $Yb_{0.98}Ca_{0.02}$ as a function of temperature measured at 35 GHz and 9·6 GHz. The curves represent the calculated g-value according to the equation for the bottleneck, including dynamic effects (after Schmidt, A 133).

Furthermore, in these alloys the E.S.R. linewidth does not show behaviour typical of magnetic ordering and again it is strongly frequency dependent. In the fuller of the two papers (A 133) the bottleneck of Eu in pure Ca is shown as an increase in $d\Delta H/dT$ with decreasing Eu concentration. The g-value was found to be equal to the free-ion value (1·993) at all concentrations. In this paper the dynamic effects were illustrated for a whole range of Ca concentrations in the YbCa host and spin-flip scattering rates for Eu and Ca were deduced. We note that the g-value quoted by Ehara and Arrott (A 69) for an alloy of 20% Eu in Yb at 295 K was of opposite sign ($g = 2·103 \pm 0·02$) to the lower concentration alloys ($< 10\%$) studied by Schäfer *et al.* reported above, although an alloy of 20% Eu in Y at 295 K had a g-value of $1·985 \pm 0·01$ (K-band).

Pop and Chechernikov (A 117) studied solid solution alloys of Gd in Y, Er and Tb mainly using high Gd concentrations. In each alloy system the Gd room temperature g-value passed through a maximum and the linewidth through a minimum for alloys containing about 20% Gd. As a function of temperature the linewidth of a Gd–Tb alloy (concentration unspecified) was found to decrease from 2·4 kG at 100°C to 1·2 kG at 20°C whilst the g-value increased from 1·87 to 2·35 in the same temperature range. Presumably, had a wider temperature range been explored, both the g-value and linewidth would have passed through minima, although the results suggest that the linewidth would have given a minimum at much lower temperatures than the g-value— contrary to the behaviour normally associated with magnetic ordering.

Recently, the Y–Gd system has been studied carefully for both high and low concentrations using single-crystal samples. Elschner and Weimann (A 71) looked at concentrations of 0·15, 0·6 and 1·0 at. % Gd in Y and compared the results with polycrystalline samples in the same concentration range. In the single crystals, evidence of fine structure was obtained from the angular dependence of the field for resonance. For the 0·6 at. % sample an isotropic g-value of $1·991 \pm 0·005$ was observed whereas a g-shift of $+0·025$ was obtained for the 0·15 at. % sample; this, coupled with a clear increase in $d\Delta H/dT$ on decreasing the Gd concentration, confirmed the presence of an E.S.R. bottleneck. This seemed to be even more marked in the single-crystal samples. In a further paper (A 158) on the same alloy system Elscher and Weimann discussed the unresolved fine structure in more detail, relating it to the theory of Barnes (1974) and Plefka (1972) and showing that the Korringa rate was large enough that the fine structure should be narrowed to a single line. The data extrapolated to the dilute limit gave $\Delta g = 0·05$ and a Korringa slope of 250 ± 75 G . K^{-1}. The axial term in the spin Hamiltonian D, was given as -140 ± 20 G. At higher concentrations (2·4 at. %) the antiferromagnetic resonance was studied in detail. Since the negative sign of D in paramagnetic resonance showed that the crystal field favours an alignment of the magnetic moments parallel to the c-axis, whilst the a–b plane was found from the angular variation of the antiferromagnetic resonance to be the easy plane for the magnetization in the ordered state, it was concluded that the anisotropy field, H_A, does not in this system have its origin in the anisotropy of the crystal field. Instead the authors suggested that the indirect exchange interaction between the Gd ions was the source of the anisotropy. The antiferromagnetic resonance was only observed at 24 and 35 GHz (at X-band, $2 H_E H_A > (w/\gamma)^2$). From the form of $(2 H_E H_A)^{1/2}$ as a function of temperature (fitted to a Brillioun function) a value of $T_N = 8·5$ K was obtained for the ordering temperature (fig. 24), in agreement with the results of other measurements on the system (70).

Bagguley et al. (A 9) made a careful and detailed study of the microwave resonance of six GdY alloys in the form of single crystals with Gd concentrations ranging from 6% to 64% Gd, measurements being made in the temperature range 4 to 293 K both at 9·5 GHz. In the 6% and 28% alloys, resonance was observed in the spiral phase with $g = 2·02 \pm 0·02$ and $1·99 \pm 0·02$ respectively compared with $1·98 \pm 0·02$ and $1·95 \pm 0·02$ in the paramagnetic regimes. Just below the Neel temperature in both of these alloys the signal intensity was a maximum, and at 9 GHz both lines passed through a minimum in linewidth close to T_N (fig. 25). At 36 GHz in the 6% alloy the linewidth continued to

Fig. 24

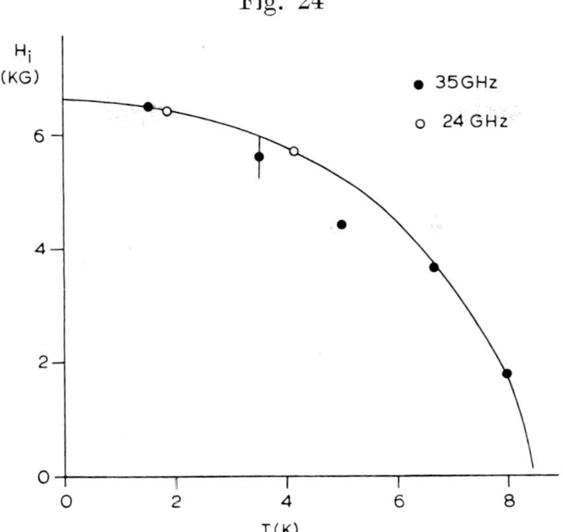

The measured inner fields $H_i = \sqrt{2H_E H_A}$, where H_E is the exchange field and H_A the anisotropy field of the antiferromagnetic Y–Gd alloy (2·4%), fitted to a Brillioun function of spin 7/2 (after Weimann and Elschner, A 158).

Fig. 25

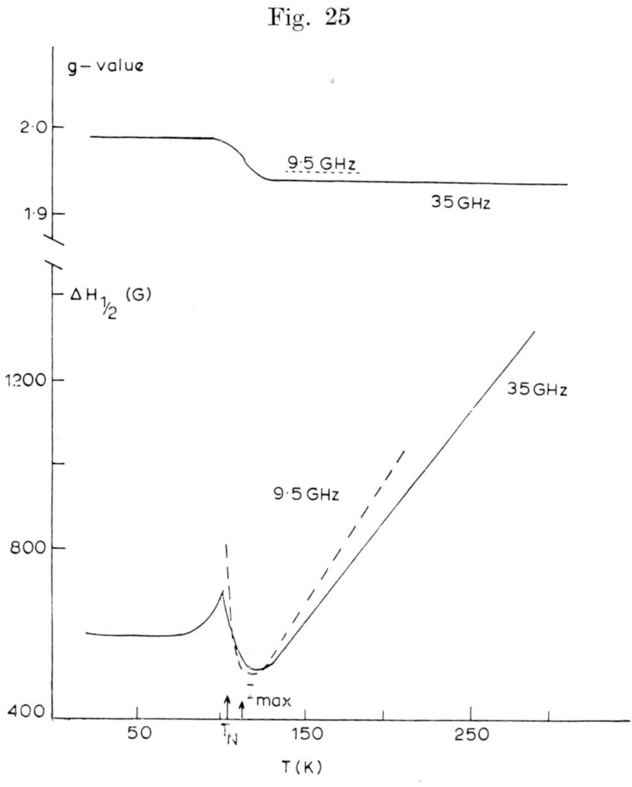

Temperature dependence of the *g*-value and linewidth for Y 28 at % Gd (Bagguley *et al.*, A 9).

narrow below T_N whilst in the 28% alloy the line broadened but then flattened out at low temperatures.

The results for the higher concentrations were more complex. For the 53% alloy, a field of 8 kG in the basal plane or 16 kG along the c-axis was sufficient to lead to a metamagnetic transition. At 9·5 GHz, the resonance line broadened in both the $1\bar{1}00$ and 0001 disks after passing through a linewidth minimum, whilst at 35 GHz a discontinuous change in linewidth and intensity occurred close to the Neel temperature when the field was aligned along the (0001)–c-axis but for field directions in the basal plane no such effects were observed and in this case the linewidth showed a maximum just below T_N. The resonance at 35 GHz was found to be isotropic in the 0001 plane and anisotropic in the 1100 plane below 220 K. Except close to T_N the (0001)–c–axis was the hard direction. For all the higher concentration alloys a g-value in the paramagnetic regime of $1·96 \pm 0·02$ was observed. Because the metamagnetic transition occurs at 4·5 kG and 3·2 kG for the 58% and 59% disks, the 35 GHz measurements showed typical ferromagnetic behaviour. In the 64% alloy the spiral phase is absent and it is ferromagnetic in zero field. At X-band for these alloys the $1\bar{1}00$ disk only gave a resonance for orientations close to the (0001)–c–axis below T_N and the linewidth in this orientation passed through a strong minimum close to the Neel point. For the 0001 disk the linewidth passed through a similar minimum but then peaked sharply just below T_N followed by a further minimum.

Full analysis of this complex resonance behaviour combined with magnetization measurements enabled Bagguley and his co-workers to elucidate the magnetic phase diagram for Y–Gd. Sarkissian and Coles (private communication) have recently shown that below about 2·5 at. % Gd there exists a regime of spin-glass-type behaviour.

Anisotropy constants were derived from the ferromagnetic resonance data. The cylindrical terms K_2, K_4 in the expansion for the anisotropy energy

$$F_A = -K_2 \sin^2 \theta + K_4 \sin^4 \theta + \ldots$$

were found to be of order 3×10^6 erg cm^{-3}, which is an order of magnitude greater than might be expected for an S-state ion. It was suggested that these large values may be due to multiple spin interactions. Once again the sign of J deduced from E.S.R. results appears to disagree with the values obtained from the magnetization data.

The results of this section are summarized in table 3.

Table 3. Summary of E.S.R. data for Gd and Eu in Y, Lu, Sc, La and Yb

Impurity	Host	Concentration at. (%)	Form of sample	Frequency	g-value	$\dfrac{d\Delta H}{dT}$	References
Eu	Y	20	Powder	K-band	$1{\cdot}985 \pm 0{\cdot}01$	—	(A 96) Ehara, Arrott
Eu	Yb	20	Powder	K-band	$2{\cdot}03 \pm 0{\cdot}02$	—	(A 69) Ehara, Arrott
Eu	Yb	1	Plates	35	$1{\cdot}970 \pm 0{\cdot}005$ at $1{\cdot}6$ K		(A 129, A 131, A 133, A 134)
		0·3					Schmidt *et al.*
					g-value and Korringa slopes frequency and temperature dependent (see text)		
		0·5					
Gd	La	13	Powder	35	$1{\cdot}995$	—	(A 82) Harris *et al.*
Gd	La	0·3	Powders	35	$2{\cdot}03 \pm 0{\cdot}02$	70	(A 35) Cottet
		1·0		35	$2{\cdot}03 \pm 0{\cdot}02$	48	
		2·0		35	$2{\cdot}03 \pm 0{\cdot}02$	26	
Gd	Lu	8·3	Powders,	35	$2{\cdot}015 \pm 0{\cdot}01$	—	(A 81) Harris *et al.*
		15	Spheres	35	$2{\cdot}010 \pm 0{\cdot}01$	—	
		20·2	and discs	35	$2{\cdot}00 \pm 0{\cdot}01$	—	
		34·6		35	$2{\cdot}010 \pm 0{\cdot}005$	—	
		47·6		35	$2{\cdot}005 \pm 0{\cdot}005$	—	
		57·2		35	$2{\cdot}004 \pm 0{\cdot}005$	—	
		69·4		35	$1{\cdot}986 \pm 0{\cdot}005$	~ 6 G . K^{-1}	
		80·6		35	$1{\cdot}983 \pm 0{\cdot}005$	—	
Gd	Lu	81·3	Powder	35	$2{\cdot}010 \pm 0{\cdot}01$	—	(A 82) Harris *et al.*
Gd	Sc	9·9	Powder	35	$1{\cdot}995$	—	(A 8) Harris *et al.*
Gd	Sc	0·3	Powders	35	$2{\cdot}00 \pm 0{\cdot}02$	32	(A 35) Cottet

Table 3 (*Continued*)

		1·0		35	$2·00 \pm 0·02$	23	
		2·0		35	$2·00 \pm 0·02$	18	
Gd	Sc	0·05	Single crystal	9	1·988–2·075 anisotropic	—	(A 127) Salamon
Gd	Y	3·7	Powders	35	2·00–2·12	—	(A 81) Harris *et al.*
		10		35	$2·12 \pm 0·05$	—	
		25		35	$2·00 \pm 0·03$	—	
		33		35	$2·00 \pm 0·03$	—	
		40		35	$1·98 \pm 0·03$	—	
		50		35	$1·94 \pm 0·03$	—	
		60		35	$1·99 \pm 0·03$	—	
		64		35	$1·98 \pm 0·03$	—	
		70		35	$1·94 \pm 0·03$	—	
		75		35	$1·94 \pm 0·03$	—	
		86·6		35	$1·96 \pm 0·03$	—	
Gd	Y	5	Powders	35	$1·993 \pm 0·005$	—	(A 82) Harris *et al.*
		10		35	$1·992 \pm 0·005$	—	
		32		35	$1·983 \pm 0·005$	—	
		47		35	$1·979 \pm 0·005$	—	
Gd	Y	25	Powders	9	2·10	—	(A 117) Pop, Chechernikov
		50		9	2·22	—	
		65		9	2·29	—	
		75		9	2·30	—	
		90		9	2·28	—	
Gd	Y	0·2	Powders	35	$2·04 \pm 0·02$	108	(A 35) Cottet
		0·6		35	$2·04 \pm 0·02$	49	
		1·0		35	$2·04 \pm 0·02$	40	
		2·0		35	$2·04 \pm 0·02$	33	

Table 3.(*Continued*)

Impurity	Host	Concentration at. (%)	Form of sample	Frequency	g-value	$\dfrac{d\Delta H}{dT}$	References
Gd	Y	0·15	Powder and single crystal	35	$2\cdot020 \pm 0\cdot005$	powder ~ 80 S.C. ~ 50	(A 71) Elschner, Weimann
		0·33	Powder	35	—	~ 48	
		0·64	Powder and single crystal	35	—	powder ~ 38 S.C. ~ 10	
		1·1	Powder and single crystal	35	$1\cdot991 \pm 0\cdot005$	~ 28	
Gd	Y	0·2	Single crystal	9·6, 24 and 35	1·99	—	(A 158) Weimann, Elschner
		2·4	Single crystal	9·6, 24 and 35	2·02	—	
Gd	Y	6	Single crystals	9·5, 24	$1\cdot98 \pm 0\cdot02$	5	(A 9) Bagguley *et al.*
		28		9·5, 35	$1\cdot95 \pm 0\cdot02$	5	
		53		9·5, 35	$1\cdot96 \pm 0\cdot02$	5·5	
		58		9·5, 35	$1\cdot96 \pm 0\cdot02$	3	
		59		9·5, 35	$1\cdot96 \pm 0\cdot02$	3	
		64		9·5, 35	$1\cdot96 \pm 0\cdot02$	3	

4.6. *Solid solution alloys of Gd and Eu($4f^7$) in 'simple' metals (Ag, Au, Al, Mg)*

The solubility (due to size factor considerations) of Gd in Cu is negligible but there exists limited solubility in the four other simple elemental hosts considered in this section.

Nakamura *et al.* (A 104) prepared a series of solid solutions of Gd in Ag with concentrations of 0·15, 0·44 and 0·5 at. % Gd and reported a linewidth linear with temperature down to the lowest temperatures obtained (4·2 K) with a slope ~ 4 G.K^{-1} for the low concentration alloy. The g-value for all the alloys was close to the Gd free-ion value and both the slope of the linewidth and this g-value increased on the addition of Al impurities. These observations were ascribed to the breaking of an E.S.R. bottleneck. The results obtained in this work agree well with the result of the Ag 0·8 at. % Gd alloy examined by Coles and Griffiths (A 79) who obtained $g = 2 \cdot 005 \pm 0 \cdot 005$ and $\Delta H = 158$ G at 4·2 K.

However, the results of Davidov *et al.* (A 57) seem to be quite different. A g-value of $2 \cdot 08 \pm 0 \cdot 01$ was found at very low Gd concentrations although a further line for Gd concentrations of around 500 p.p.m. was observed with a g-value close to 2. This latter line was attributed to clusters. No angular dependence of the g-value in single-crystal samples was observed and it was noted that the addition of 1% Au did not alter the observed g-value. It was thus concluded by Davidov *et al.*, in conflict with the conclusions of Nakamura *et al.*, that Gd in Ag does *not* display a resonance bottleneck.

One notable and interesting attempt to study much higher concentrations of Gd in Ag was carried out by Charles *et al.* (A 22). In these experiments E.S.R. of films of Gd in Ag, vapour quenched onto substrates at various temperatures was studied. X-ray analysis suggested that alloys deposited at low temperatures with Gd concentrations in excess of 53 at. % and those formed at room temperature with 89 and 94% Gd were amorphous. The E.S.R. linewidths were used to obtain estimates of the Curie temperatures of the alloys—the temperature of the linewidth minimum being taken for this purpose. It was found that for low temperature deposited alloys a Curie temperature against composition plot gave a minimum of 40 K at 88 at. % Gd rising to 360 K for 30 at. %, whereas for the room temperature deposited alloys the values of T_c were all close to 110 K for concentrations between 30 and 80 at. % Gd. The g-values all fell within the range $1 \cdot 950 \pm 0 \cdot 010$ to $2 \cdot 020 + 0 \cdot 020$ but with no clear concentration dependence. We note that as in the work of Taylor and Coles (A 151) a number of these concentrated alloys gave large negative residual linewidths.

It was in the alloy system Au–Gd that the first resolved fine structure of a localized moment in a metal was reported by Chock *et al.* (A 24) using single crystals containing 1000, 300 and 150 p.p.m. Gd in Au. At 4·2 K they observed a single line at all angles, but with considerable angular variation of the linewidth. At 1·4 K, however, they resolved fine structure at certain angles. The resonance position of the 'line' at 1·4 K also showed an angular variation of 30 ± 5 G. For minimum linewidth a g-value of $2 \cdot 05 \pm 0 \cdot 01$, equal to that obtained from powder samples of the same concentration, was observed. The linewidth at its angular minimum was fitted to a form $A(c) + BT$, as a function of temperature, where $B = 8$ G . K^{-1} (as in powder samples) though the residual width, A, for a 150 p.p.m. sample at 17 ± 3 G was four times smaller than the value for the same Gd concentration powder sample. Addition of Pt impurities decreased B, thus suggesting that contrary to some reports (see above) on Ag-based alloys the system is not bottlenecked. It was the fact that the observed resonance in this system did not display that character expected for resolution of the seven Gd fine-structure lines that led to the development of the narrowing mechanisms discussed in § 2.

The sign of the g-value of dilute solutions of Gd (and Eu) in Mg has been the subject of some disagreement. Ehara (A 68) measured an alloy of 0·5 at. % Gd in Mg and obtained a g-value of $2·03 \pm 0·005$ with a linewidth of 260 ± 30 G at 4·2 K and interpreted the positive shift as evidence in explaining the logarithmic increase of the incremental resistivity with increasing temperature as a Kondo effect with positive J. Williams (1972), however, re-emphasized the dangers of comparing the sign of J obtained in different measurements. Tao *et al.* (A 147) measured alloys of 500 p.p.m. Gd in of single of Mg at 1·4 K and 4·2 K at both X-band and Q-band. They reported strongly anisotropic g-values which were temperature dependent and *negative* at 1·4 K for all angles. As is the case

Fig. 26

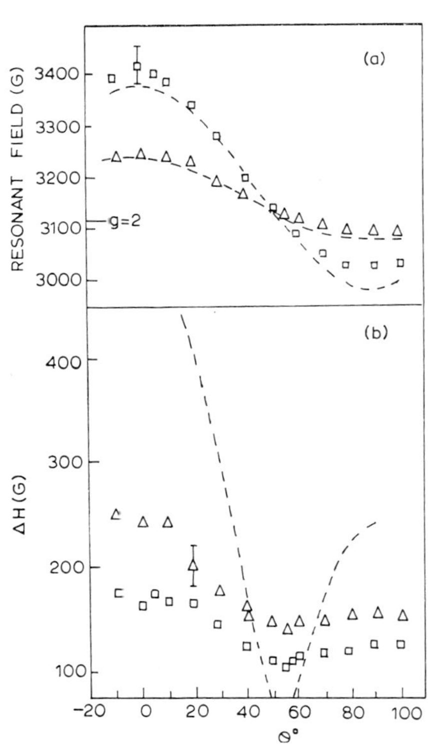

(a) The field for resonance as a function of the angle θ between the magnetic field and the c-axis for Mg : Gd (500 p.p.m.) at a frequency of 8·7 GHz. The squares indicate measurements at 1·4 K; the triangles 4·2 K. The dashed lines are the first moments, as calculated using values of $D = 155$ G and $g = 1·98$ at the respective temperatures. (b) The linewidth as a function of angle for the same sample. The dashed line is the square root of the second moment as calculated using the same values of D and g (Tao *et al.*, A 147).

Fig. 27

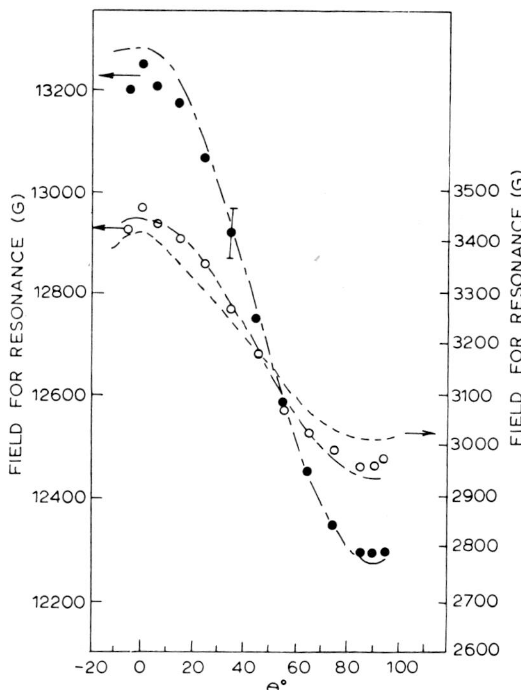

The field for resonance as a function of the angle θ between the magnetic field and the
c-axis for Mg : Gd (500 p.p.m.) at a frequency of 35 GHz.   The solid circles
indicate measurements at 1·6 K; the open circles 4·2 K.   The dash–dot lines
are the first moments as calculated using $D = 155$ G and $g = 1·98$.   For
comparison the angular dependence of the field for resonance at 8·7 GHz at
1·4 K (dashed line) is shown (Tao et al., A 147).

Fig. 28

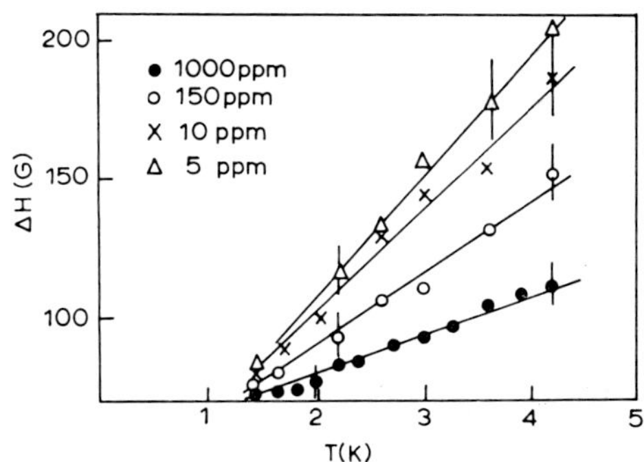

E.P.R. linewidth of Gd in Al : Gd dilute alloys as a function of temperature.   Samples
were melted many times, starting from a master sample and diluted step by
step.   The data shows clearly the bottleneck in this system (Rettori et al.,
A 122).

Table 4. Summary of E.S.R. data for Gd in Ag and Au

Impurity	Host	Concentration (at. %)	Frequency	Form of sample	g-value	$\dfrac{d\Delta H}{dT}$	References
Gd	Ag	3	—	Powder	$1{\cdot}995 \pm 0{\cdot}007$	—	(A 115) Peter *et al.*
Gd	Ag	0·8	9	Powder	$2{\cdot}005 \pm 0{\cdot}005$	4·4	(A 79) Griffiths, Coles
Gd	Ag	0·15	9·2	Powder	1·987	3·8	
		0·44	9·2		1·987	3·2	
		0·51	9·2			3·2	
Gd	Ag	500 p.p.m.	—	Powders and single crystals	$2{\cdot}08 \pm 0{\cdot}01$ (two lines)	—	(A 57) Davidov *et al.*
		25 p.p.m.	—		$2{\cdot}08 \pm 0{\cdot}01$	19·7	
		50 p.p.m.	—		$2{\cdot}08 \pm 0{\cdot}01$	19·7	
Gd	Ag	96	9	Thin film deposited at 4·2 K except where indicated	$1{\cdot}970 \pm 0{\cdot}010$	$5{\cdot}07 \pm 0{\cdot}39$	(A 22) Charles *et al.*
		94	9		$1{\cdot}980 \pm 0{\cdot}010$	$5{\cdot}21 \pm 1{\cdot}24$	
		90	9		$1{\cdot}980 \pm 0{\cdot}010$	$2{\cdot}03 \pm 0{\cdot}51$	
		86	9		$2{\cdot}010 \pm 0{\cdot}010$	$1{\cdot}81 \pm 0{\cdot}27$	
		85	9	(78 K)	$1{\cdot}990 \pm 0{\cdot}010$	$1{\cdot}40 \pm 0{\cdot}65$	
		84	9		$1{\cdot}960 \pm 0{\cdot}010$	$1{\cdot}50 \pm 0{\cdot}34$	
		79	9		$1{\cdot}960 \pm 0{\cdot}010$	$1{\cdot}57 \pm 0{\cdot}40$	
		76	9		$1{\cdot}960 \pm 0{\cdot}010$	—	
		54	9		$2{\cdot}020 \pm 0{\cdot}020$	$1{\cdot}60 \pm 0{\cdot}20$	
Gd	Ag	46	9	Thin films (78 K)	$1{\cdot}980 \pm 0{\cdot}010$	—	

Table 4 (*Continued*)

		23	9	(78 K)	$1 \cdot 980 \pm 0 \cdot 010$	—	
Gd	Ag	94	9	Thin films deposited at room temp.	$1 \cdot 960 \pm 0 \cdot 010$	$9 \cdot 35 \pm 3 \cdot 14$	(A 22) Charles *et al.*
		88	9		$1 \cdot 970 \pm 0 \cdot 010$	$3 \cdot 82 \pm 1 \cdot 04$	
		80	9		$1 \cdot 965 \pm 0 \cdot 015$	$1 \cdot 87 \pm 0 \cdot 25$	
		66	9		$1 \cdot 950 \pm 0 \cdot 010$	$2 \cdot 74 \pm 0 \cdot 54$	
		59	9		$1 \cdot 950 \pm 0 \cdot 010$	$1 \cdot 70 \pm 0 \cdot 27$	
Gd	Au	150 p.p.m.	8·7	Single crystal	$2 \cdot 05 \pm 0 \cdot 01$	6 ± 1	(A 24) Chock *et al.*
		300 p.p.m.	8·7	Single crystal	$2 \cdot 05 \pm 0 \cdot 01$	7 ± 2	
		35 p.p.m.	8·7	Powder	$2 \cdot 05 \pm 0 \cdot 01$	8 ± 1	
		75 p.p.m.	8·7	Powder	$2 \cdot 05 \pm 0 \cdot 01$	7 ± 1	
		150 p.p.m.	8·7	Powder	$2 \cdot 05 \pm 0 \cdot 01$	7 ± 2	
		500 p.p.m.	8·7	Powder	$2 \cdot 05 \pm 0 \cdot 01$	6 ± 2	

of Salamon's results for Gd in Sc (A 127) a minimum in the linewidth was observed for an angle of about 60° between the applied d.c. field and the *c*-axis. Tao and his co-workers proposed that both sets of results could be best described in terms of unresolved fine structure (see figs. 26 and 27).

The sign of the *g*-value for Gd in Mg is further confused by the report of Burr and Orbach (A 10) of a positive *g*-value for both Gd and Eu in Mg, although they declined to give a precise value "due to deviations from Dysonian Lineshape".

Fig. 29

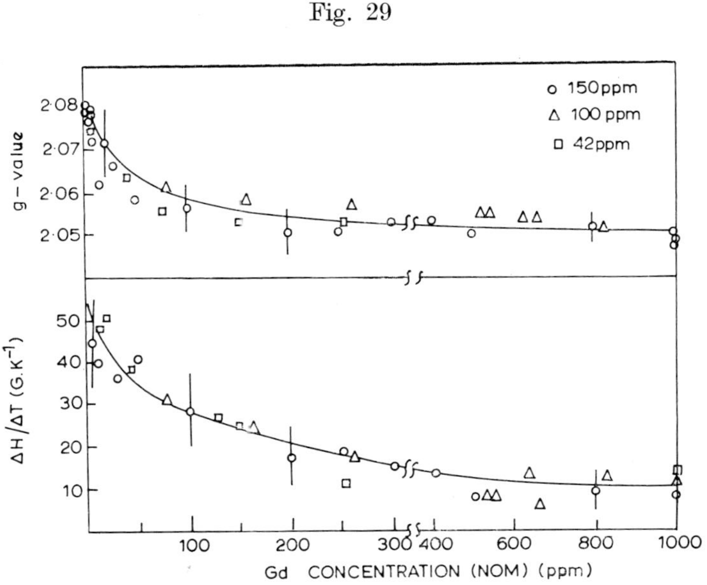

The *g*-value and the thermal broadening versus concentration for all the Al : Gd samples measured by Rettori *et al.* (A 122). The solid lines represent the theoretical fit.

Finally in this section we mention the results of Rettori *et al.* (A 122) for Gd in Al. A strong bottleneck was observed (see figs. 28 and 29) with a Korringa slope of 8 G.K⁻¹ for a 1000 p.p.m. sample and ~ 55 G.K⁻¹ in the dilute limit, the *g*-values for the same concentrations increasing from 2·05 to 2·08. The existence of a bottleneck was confirmed by the addition of Ag, Au and Pt, 4000 p.p.m. of Au, for example, increasing $d\Delta H/dT$ for the alloy containing 1000 p.p.m. Gd to 30 G.K⁻¹. Values were obtained for the spin–lattice relaxation rate of the conduction electrons from both the Gd and third elements. Rettori and his co-workers considered possible mechanisms that might lead to the apparent flattening out of the *g*-values in the strongly bottlenecked limit to the positive *g*-shift of 0·05 ± 0·01. One possible explanation envisaged a contribution of this sign due to the coupling between the local moment and enhanced 5d screening electrons.

The results of E.S.R. in Ag and Au hosts are summarized in table 4.

4.7. *Gd in other hosts*

In this section brief mention will be made of a few results which do not fall appropriately in any of the above categories of host.

As part of their study of various rare earths in Rh, Davidov *et al.* (A 53, A 42) examined the E.S.R. of a 110 p.p.m. Gd in Rh powder and obtained a g-value of $1 \cdot 98 \pm 0 \cdot 01$ and a Korringa slope of $7 \pm 2\,\mathrm{G}.\mathrm{K}^{-1}$. A study of rare earths in Th gave for both splat-cooled and powders alloys containing between 95 and 1300 p.p.m. Gd g-values of $2 \cdot 01 \pm 0 \cdot 01$ and Korringa slopes of $31 \pm 5\,\mathrm{G}.\mathrm{K}^{-1}$.

To close this section on s-state ions in various hosts we mention the measurements of Cottet *et al.* (A 34) on 2% Gd in Rh–Ni alloys. Measurements made at 85 K to avoid the complicating effect of a Ni ferromagnetic resonance observed at 4·2K show a g-value which increases from 1·89 for an alloy of 35 at. % Rh in Ni to about 1·97 for Gd in pure Rh. At the same time the corresponding linewidths were observed to decrease progressively from $\sim 1200\,\mathrm{G}$ to $\sim 700\,\mathrm{G}$ in pure Rh. The g-shift results clearly show the increasing d-character at the Fermi surface as Ni is added to Rh.

4.8. *Mn in Cu, Ag and Au*

The first observation of an impurity resonance in a metal was made by Owen *et al.* (A 110, A 111) on alloys of Mn in Cu. They were able to study alloys containing from 0·07% Mn up to 11·1% Mn at 2 K, 4 K, 77 K and 295 K, and also looked at one alloy of 4·2% Mn in Ag. The E.S.R. linewidths increased with temperature, and g-values in the range $g = 1 \cdot 99$ to $2 \cdot 01$ were obtained after correcting for lineshape. It was these values, close to the free-ion values, which surprised Owen *et al.* and provoked the important concept of bottlenecking developed by Hasegawa. At low temperatures in the alloys containing more than about 1 at. % Mn in Cu, they observed from susceptibility measurements what appeared to be antiferromagnetic ordering at temperatures increasing from about 12 K for the 1·4 at. % alloy to about 100 K for the 11·1%. In addition, they observed a spontaneous magnetization at 4K from the higher concentration alloys, the value of which appeared to be a function of the field in which the alloy was cooled. Below the ordering temperature, the E.S.R. line was observed to move to lower fields, the largest shift being obtained in the highest concentrations (at 4·2 K, the 5·6% Mn alloy was shifted by 1170 G; the 1·4% alloy by 190 G). The shifts were found to be well described by the relation (see § 2)

$$H_{\mathrm{RES}} = (H_0{}^2 - H_{\mathrm{c}}{}^2)^{1/2}$$
$$H_{\mathrm{c}} = (2H_{\mathrm{E}}H_{\mathrm{A}})^{1/2}$$

which is of the same form as the field for resonance condition for an antiferromagnet with H perpendicular to the anisotropy axis.

Field cooling was also found to have a strong effect on the E.S.R. results. Cooling the 5·6 at. % alloy from 77 K to 4 K in a field of 5 kG shifted the line back to higher fields by 670 G, compared to a measurement taken at 4·2 K after cooling in zero field. Hysteresis effects and gradual field-dependent shifts were also found when the alloy was held at 4 K and the field varied. Measurements on a single crystal of 5·6 at. % Mn in Cu showed the line to be isotropic to within $\pm 30\,\mathrm{G}$ in the (111) and (110) planes.

Alloys of 15, 20, 25 and 30 at. % Mn in Cu were investigated by Street (A 144). At high temperatures, g-values of 2·0 were obtained and in each case line broadening was observed as the temperature was decreased, the line disappearing at a characteristic temperature, which was found to be greater than that at which the susceptibility reached a peak value by an amount increasing rapidly with Mn concentration (for a 30% alloy ~ 200 K). At or near the temperature of the susceptibility maximum a symmetrical narrow line was observed ($\Delta H \sim 100$ G, $g = 2$) which on occasions was observed simultaneously with the broader line. This was interpreted as being due to a ferromagnetic resonance. Street concluded that the observation of a broadening resonance line above the susceptibility maximum was due to short-range order and it was suggested that the ferromagnetic line originated from small groups of ions with a net magnetic moment.

Shaltiel and Wernick (A 133) observed the resonance of 2% Mn in Au, Cu and Ag and confirmed the previously obtained g-values, obtaining $g = 2·005 \pm 0·0005$ for Mn in Au. The line in this case was much broader at 20 K (~ 725 G) than the Cu–Mn (~ 140 G) or Ag–Mn (~ 120 G) lines. The single-crystal result of Owen *et al.* was also confirmed by Geschwind *et al.* (A 75) on a 0·1 at. % Cu–Mn single crystal. They observed a line which was only 40 G broad at low temperatures and were unable to find evidence of either hyperfine structure or anisotropy in the g-value to one part in four thousand.

Griffiths (A 78) using a continuously variable temperature spectrometer carefully investigated the behaviour of alloys containing 4 at. % and 15 at. % Mn in Cu as a function of their magnetic history. In the 4 at. % alloy cooled in zero field (fig. 30), a well-defined maximum in the plot of signal amplitude ($A + B$) against temperature was observed at 43 ± 2 K. At about the same

Fig. 30

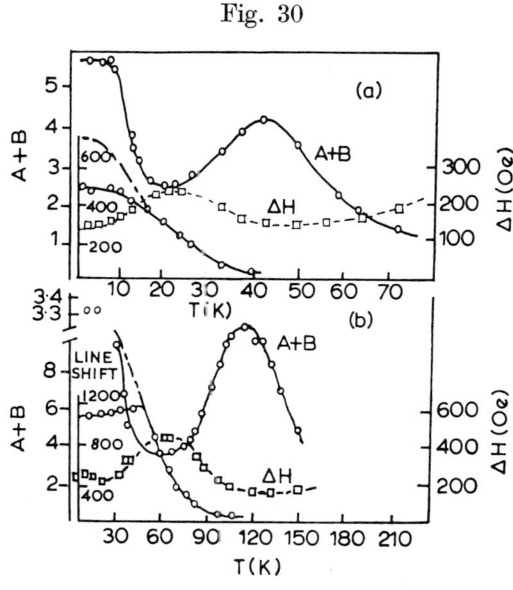

Cooling in a 5 kOe field. (*a*) Cu–4 at % Mn; (*b*) Cu–15 at % Mn. Other details as in fig. 30. The chain line indicates the line shift that was observed with zero field cooling (Griffiths, A 78).

Fig. 31

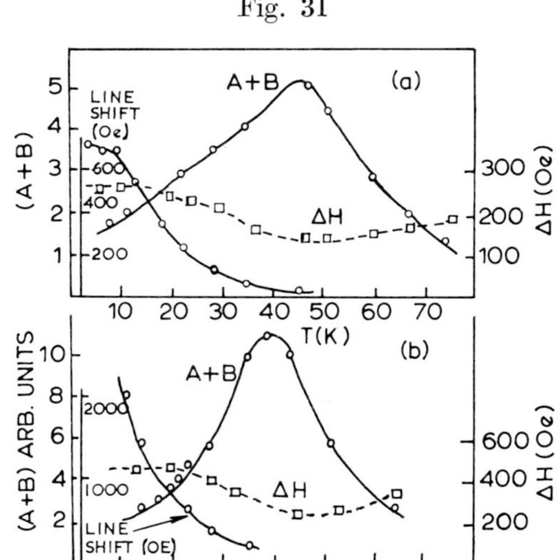

Cooling in zero field: (a) Cu–4 at % Mn; (b) Cu–15 at % Mn. The spin resonance signal amplitude $A+B$, the linewidth ΔH and line shift plotted against temperature (after Griffiths, A 78).

temperature (50 K) the linewidth showed a weak minimum and a line shift became apparent. The more concentrated alloy showed similar behaviour, the corresponding temperatures being 115 ± 3 K for $(A + B)$ and 135 K for linewidth. The g-values and linewidths obtained by Griffiths agreed well with those of Owen *et al.*, except for the room temperature linewidth of the 4 at. % alloy which Griffiths found to be considerably narrower than Owen's value, although both samples were annealed. Griffiths also found a smaller shift in resonant frequency for his 4 at. % alloy (700 ± 20 G). Cooling the alloys in a field of 5 kG radically altered the behaviour on a subsequent warm-up run (fig. 31). The temperature of the amplitude maximum was the same as the zero-field-cooled value but showed, in addition, a pronounced minimum at 21 K and 60 K for the 4% and 15% alloys, respectively. There was a maximum in the linewidth at these temperatures and a reduction in the shift in resonant field at low temperatures; the 4 at. % alloy was, for example, only shifted 460 ± 20 G from $g = 2$ at 4·2 K. By plotting the parameter $(A + B)(\Delta H^2)$ (proportional to the total intensity and thus the magnetic susceptibility) against temperature, it was evident that a $1/T$ dependence of the signal intensity did not hold. The form of the curves both for zero and field-cooled samples was similar; using the expression $H^2_{\mathrm{RES}} = H_0^2 - H_c^2$ used by Owen *et al.*, Griffiths found a linear relation between H_c and T for the zero-field data and noticed that the extrapolated value of T corresponding to $H_c = 0$ was in both cases close to the temperature of the observed amplitude maximum. Griffiths pointed out the fact that whilst these characteristic temperatures agreed well with those obtained by static susceptibility measurements, they are much lower than anomalies found in the resistance data for Cu–Mn alloy. The E.S.R. results were thought to support Kouvel's (1961) theory of ferromagnetic/antiferromagnetic interactions and the characteristic temperatures to correspond to the onset of ferromagnetic interactions. Cu–Mn is now considered to be an archetypal spin-glass (or mictomagnetic) system and it is apparent that much is to be learnt from E.S.R. measurements about the excitations in these materials.

Gossard *et al.* (A 76) studied the temperature dependence of the E.S.R. linewidth of Cu–Mn alloys containing 0·3 and 1·5 at % Mn from 1·4 K to 240 K and the effects of adding from 0·01% to 0·15 at. % of Ti, Fe, Co and Ni impurities. They estimated, on the basis of the Korringa formula, that the slope of the linewidth *should* be ~ 2000 G . K^{-1} compared with an observed value some 200 times less for a 0·3 at. % alloy and 800 times less for a 1·5 at. % alloy. Addition of Ti and Ni was found to break this bottleneck and the increase in slope of the linewidth was found to be roughly proportional to the impurity concentration. When Fe and Co were added, the broadening was an order of magnitude greater than for the other impurities and this was attributed to the extra relaxation mechanism involving the transfer of spin energy from the Mn to the Fe and Co spins. The spin-flip scattering cross sections were thus deduced for the first time from experiments of this type and these cross sections together with those of Al and Mg were tabulated.

Okuda and Date (A 107) added Ni and Al to a Cu 5% Mn alloy and found that the slope of the linewidth was five times greater for Ni. They found that in the antiferromagnetic regime for a Cu 2% Mn alloy that addition of Ni increased the lineshift whilst Al had little effect upon it. In a fuller paper (A 109), the same authors presented the results of detailed measurements on alloys of 1·4, 2·0, 3·0 and 4·0 at. % Mn in Cu below 77 K and the results were in good agreement with previous work. It was observed that the concentration dependence of the lineshift at 4·2 K was linear with a slope of 140 G.K^{-1} at X-band. The authors claimed that the anisotropy field in Cu–Mn originated from the ^{55}Mn nuclear spin and ENDOR experiments performed on Ag–Mn alloys were claimed to support this view. The previous experiments on the addition of bottleneck-breaking elements were extended, Co, Fe, Pd, Ti and Zn being tried. The first three elements behaved like Ni in the antiferromagnetic regime, whilst Ti and Zn were like Al.

McElroy and Heeger (A 94) observed that residual linewidth in the Cu–Mn system were unchanged for Mn concentration in the range 0·1 to 1·5 at. % and this ruled out Mn–Mn interactions from being the main contribution to the residual width. Addition of non-magnetic impurities such as Si or Al to a 0·1% Mn in Cu alloy was found to increase the residual width by 72 G per at. % Si and 38 G per at.% Al. They attempted to analyse this behaviour in terms of a model invoking the admixture of conduction band wave functions into the localized 3d impurity function. This mechanism has now been shown to be wrong (Orbach *et al.*, 1973). This conclusion would appear also to invalidate the claim of Oda and Asayama (A 106) from their results of linewidth against temperature and residual linewidth against concentration for dilute Cu–Mn alloys doped with Fe, Ni and Co, that Co in Cu forms an s–d quasi-bound state with high Kondo temperature.

Nakamura and Kinoshita (A 101, A 102) investigated a number of Cu–Mn alloys with concentrations from 0·3% to 2% and found that below about 1% Mn, there was a rapid increase in linewidth for decreasing concentration, with a slope rather less than inversely proportional to concentration. They observed that the residual linewidth was concentration independent for less than 1% Mn but *decreased* above this concentration. All their alloys gave $g = 2 \cdot 01 \pm 0 \cdot 01$ in the paramagnetic regime.

The importance of the metallurgical treatment of Cu–Mn alloys was highlighted by the work of Miyako *et al.* (A 96). They annealed their samples in a vacuum of 10^{-8} torr at $850\,^{\circ}$C and claimed that the 'cleanliness' of the anneal had a pronounced effect on subsequent E.S.R. behaviour. They found, for example, that use of a vacuum of only 10^{-6} torr doubled the linewidth of a 4·3 at. % Mn in Cu alloy at room temperature and that, bringing the vacuum up to 10^{-3} torr, pushed the linewidth up by a further factor of five. Contrary to the work of Owen *et al.* (A 111) and Griffiths (A 78), they claimed to find pronounced maxima in linewidth for high vacuum annealed samples at temperatures corresponding to the temperatures of the anomalies in the electrical resistivity (~ 100 K for 4·3% and 40 K for 2·0%).

One extremely interesting experiment which, to the present author's knowledge, has not been attempted elsewhere, was performed by Nagashima and Abe (A 99). They claim to have been able to open the E.S.R. bottleneck of thin films of Cu–Mn and Ag–Mn on glass plates by increasing the probability of a spin disorientation at the sample surface as the film thickness was reduced.

Although the resonance of Cr in Cu has not been observed directly by the reflection technique, Hirst *et al.* (A 84) were able to deduce information about the system by observing the resonance of Cu–Mn as modified by Cr impurities. They derived equations for the interconnected bottleneck of two impurities and obtained the relevant criteria for bottlenecking in such a case: The g-values of their Cu–Mn alloys were in good agreement ($2 \cdot 010 \pm 0 \cdot 005$) with previous values and addition of Cr was found not to alter this value for measurements at both X-band and Q-band frequences. It was found from the increase in $d\Delta H/dT$ on the addition of Cr impurities, that contrary to the cases of Fe and Co in Cu, a double bottleneck operates for Cr in Cu–Mn. The same conclusion was reached from analysis of the residual linewidths in the different systems. A g-value of $2 \cdot 01 \pm 0 \cdot 03$ was deduced for the Cr ion.

The progress of experimental work on the Ag–Mn systems has in many ways run parallel to that of Cu–Mn but has not attracted the same amount of attention. Following the single result of Owen *et al.* (A 110) on an Ag–Mn alloy, Gossard *et al.* (A 77) studied resonance in powdered stress annealed samples and on etched slabs. They found that the linewidth in their powder samples increased, as a function of temperature, at a faster rate than the slab. In a slab containing 0·26% Mn, the linewidth broadened at 10 G.K^{-1} whilst in a 1·3% alloy the corresponding rate was 3 G.K^{-1} and the authors concluded both from the absolute values and from the concentration dependence that the system, like Cu–Mn, was bottlenecked. Both alloys furthermore gave $g \sim 2$ and it was found that addition of a second species of impurity increased the slope of the linewidth. The spin-flip scattering cross sections for Zn, Ga, Cd, Pt and Au were evaluated. Elliston (A 70) showed that Ag–Mn alloys behave similarly to Cu–Mn when field-cooled and that 'characteristic' temperatures are again lower than the values obtained from resistivity measurements.

A major study of dilute Ag : Mn solid solutions has recently been carried out by Davidov *et al.* (A 54) at X-band in the liquid helium temperature range with Mn concentration of from 50 to 1500 p.p.m. The bottlenecking in the system was again clearly shown but addition of Au and Sb impurities were found to break almost totally the bottleneck. These results lead to values of $\Delta g = -0.035 \pm 0.01$ (corresponding to $J(0) = -0.25 \pm 0.07\,\text{eV}$ and $d\Delta H/dT = 45 + 10\,\text{G.K}^{-1}$ (corresponding to $\langle J_{\text{eff}}(q)^2 \rangle = 0.12 \pm 0.03\,\text{eV}^2$). In concentrations greater than 250 p.p.m. Mn, the g-value was found to be temperature dependent and this was attributed to ordering effects. Sb was found to be an excellent spin-flip scatterer in that it satisfied the criteria of having both a high cross section ($7 \pm 3 \times 10^8\,\text{sec}^{-1}$/p.p.m. Sb, almost twice the value for Au), and it led to a relatively small increase in the residual linewidth for an increase in concentration. The A/B ratio of the alloys was found to be ~ 6 at extremely low Mn concentrations and about the same value for concentrations above roughly 1000 p.p.m., reaching a peak value of 12 in a 500 p.p.m. alloy.

The g-shift and thermal broadening were analysed in terms of a set of partial wave amplitudes. Assuming again that $J_{\text{at}}(0) > J_{\text{at}}(1) > J_{\text{at}}(2)$ and that in the case of a 3d transition ion impurity $|J_{\text{c\,m}}(2)| \gg |J_{\text{c\,m}}(0)|$ with $J_{\text{cm}}(1)$ and $J_{\text{c\,m}}(3)$ negligible and including the effect of exchange enhancement, two relations between the E.S.R. parameters and the $J^{(\text{L})}$ were obtained. A third relation was obtained by using the expression for $J_{\text{eff}}(q)$ from the term in the electrical resistivity and this expression was maximized (negatively) to give the value closest to that obtained for $\langle J_{\text{eff}}(q)^3 (1 - \cos\theta) \rangle$ by Jha and Jericho (1971) ($-0.0096\,\text{eV}^3$). The closest value that could in fact be obtained was a factor of four smaller ($-0.0023\,\text{eV}^3$). The values of $J^{(0)} = 0.094\,\text{eV}$, $J^{(1)} = 0.06\,\text{eV}$ and $J^{(2)} = -0.081\,\text{eV}$ used to obtain this best value, themselves yielded a value of $J_{\text{eff}}(\text{RKKY}) = -0.80\,\text{eV}$ for the contribution to N.M.R. line broadening from the conduction electron polarization induced by the magnetic impurity—a value 25% smaller than that obtained from experiment. The values of these exchange partial wave amplitudes were shown to be roughly consistent with the magnitude of the shift from insulator values of the ^{55}Mn hyperfine field in Ag–Mn. It was concluded that the d-wave amplitude is negative and smaller in magnitude than for Mn in Cu and argued that the Kondo temperature deduced from the $J^{(\text{L})}$ would be $\sim 10^{-29}$ K for Ag–Mn (cf 2×10^{-3} in Cu–Mn).

The residual linewidth was found to decrease with increasing Mn concentration (~ 60 G for 250 p.p.m. to ~ 35 G for $C > 500$ p.p.m.) and an attempt was made to understand this and the anomalous line-shape ratio (described above) by performing a line-shape analysis similar to that of Pifer and Longo (1971) starting with seven Hasegawa-type equations of motion—six for the presumably almost totally exchange narrowed hyperfine lines ($I = 5/2$ for ^{55}Mn) and one for the conduction electron magnetization. A rate $1/T_{\text{dL}}$ was assumed to be equal to the residual width for each hyperfine line. This analysis generated a concentration dependence of the residual width in fair agreement with the experimental results and with those of Schultz *et al.* (75) measured in transmission. It did not, however, manage to explain the experimentally observed variations in A/B ratio.

4.9. *Mn in Pd*

The first detailed measurements on this system were those of Shaltiel and Wernick (A 138). They found $g \sim 2.10$ for 2 at. % sample at 4.2 K and 20 K and a Korringa slope ~ 50 G.K^{-1}. When 1% Pr was added to the alloy, the

g-value increased to 2·195 whilst addition of 0·2% Ho decreased it to 2·055. This indicated that, as in the case of Pd–Gd, the light rare earths increase the g-value whilst addition of heavy rare earths decrease it. Comparison of the values of J for the two systems showed that whilst opposite in sign their magnitudes are in fairly good agreement. The broadening of the E.S.R. line on adding the rare earth was much greater for Pd–Mn than for Pd–Gd. These measurements were extended by Cottet (A 35), who measured the temperature dependence in the paramagnetic regime and the g-values of alloys containing 0·23, 0·5, 1·0, 1·5 and 2·0% Mn. A slope of 76 G.K^{-1} was measured for the 2 at. % alloy and for the lowest concentration this had increased to 90 G.K^{-1}. A g-value of 2·12 ± 0·01 was found for the 1·5% alloy at 5 K and a value of 2·17 for the 0·23% alloy. Large positive g-shifts were also found recently by Alquié *et al.* (A 2) with a concentration-independent g-value up to 2% Mn of

Fig. 32

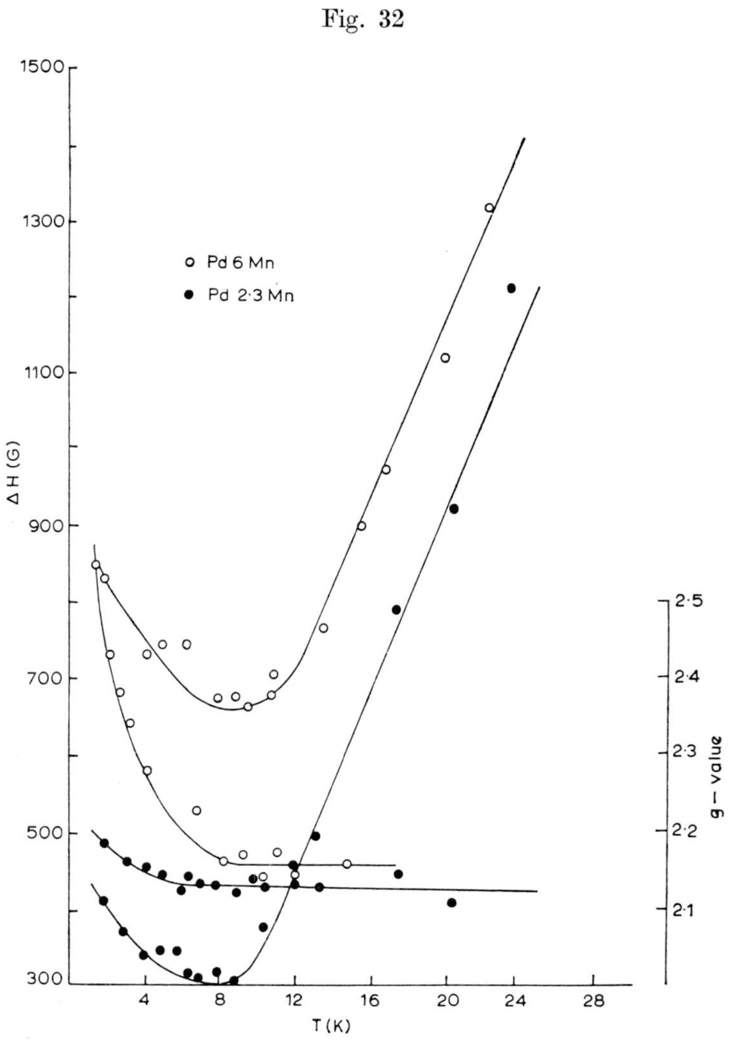

Linewidth and g-value versus temperature for two Pd–Mn alloys (Coles *et al.*, A 29).

2·12 and their values of the Korringa slope ($\sim 55\,\mathrm{G.K^{-1}}$) were in good agreement with the values of Shaltiel and Wernick. These workers also hydrogenated their alloys and as in Pd–Gd, the g-value moved back to $g = 2$, as the d-character was quenched. Values of $g = 2\cdot005 \pm 0\cdot005$ and $d\Delta H/dT = 30\,\mathrm{G.K^{-1}}$ were obtained for hydrogenated alloys of 0·1 at. % to 10 at. % Mn in Pd. In plots of the temperature dependence of the linewidth, an up-turn was found at temperatures well above the ordering temperatures (for a 10% alloy, $T_{\Delta H\,\mathrm{min}} = 20\cdot5$ whilst $T_{\mathrm{c}} = 6\cdot5\,\mathrm{K}$). Alquié and his co-workers suggested that there may be strong short-range order effects above the ordering temperatures in these alloys.

The ordering of Pd–Mn alloys was studied in detail by Coles *et al.* (A 150) showed that for concentrations up to about 3% Mn the Curie temperature rises with increasing Mn concentration (to $\sim 7\,\mathrm{K}$ for 2·5%) but then drops again, resulting in a closure of the ferromagnetic phase field at $\sim 5\%$ Mn. At high concentrations spin-glass ordering is observed with a region between the ferro-magnetic and spin-glass phase fields where there appears to be extensive short-range magnetic order. This is clearly shown by the E.S.R. results in the system (fig. 32). Line broadening sets in at temperatures close to the Curie temperature for the ferromagnetic alloys, but at higher concentrations the values of $T_{\Delta H\,\mathrm{min}}$ do not fall, but flatten out at 7–9 K. Coles *et al.* obtained g-values for concentrations of 0·6 to 7·0% Mn in Pd (above which the line became too

Table 5. Summary of E.S.R. data for Mn in Pd

Reference	Conc. (at. %)	Frequency GHz	g-value	$\dfrac{d\Delta H}{dT}$ GK^{-1}
(A 140) Shaltiel *et al.*	2	9	2·08	—
(A 35) Cottet	0·23	35	2·18	—
	0·5		2·15	95
	1·0		2·12	80
	1·5		2·12	80
	2·0		2·12	70
(A 138) Shaltiel, Wernick	2	50	2·105	—
(A 2) Alquié *et al.*	0·1	10·9	2·12	55
	1		2·12	—
	2		2·12	40
Coles *et al.*	0·6	9	2·15 ± 0·02	50
	0·9		2·14 ± 0·01	62
	2·3		2·13 ± 0·01	55
	2·5		2·15 ± 0·01	57
	3·75		2·14 ± 0·02	40
	4·6		2·16 ± 0·02	47
	5·5		2·16 ± 0·02	58
	6·0		2·16 ± 0·02	52
	7·0		2·17 ± 0·03	—

broad and weak to observe) of from $2 \cdot 13 \pm 0 \cdot 01$ to $2 \cdot 17 \pm 0 \cdot 02$ with no clear concentration dependence and values of $d\Delta H/dT$ from 40 to $62 \, \text{G.K}^{-1}$. They showed by the addition of Au and Pt impurities that the system is not bottle-necked. In the spin-glass region no field-cooling effects of the type found for Cu–Mn and Ag–Mn alloys were observed.

The collected results for the E.S.R. results in the Pd–Mn alloy system are shown in table 5.

4.10. *Mn in Mg and Zn*

In their classic paper on Cu–Mn, Owen *et al.* (A 110) performed a single experiment on a $0 \cdot 67 \%$ Mn in Mg alloy and obtained a g-value of $2 \cdot 02$ with linewidth of 270 G at $4 \cdot 2$ K, which showed some signs of Cu–Mn–type ' antiferro-magnetic' behaviour below 4 K. Single-crystal measurements in this regime showed some anisotropy. A more thorough study of E.S.R. in a $0 \cdot 6$ at. % Mn in Mg alloy was carried out by Collings and Hedgcock (A 30). They reported $g = 2 \cdot 01 \pm 0 \cdot 01$ from 70 K down to about 6 K and a shift in the resonant field due to ordering of 300 G at 2 K, which is roughly double the value reported by Owen *et al.* A temperature-independent linewidth of 300 G was found for temperatures up to 20 K but above this the linewidth broadened with a Korringa slope of $6 \, \text{G.K}^{-1}$. A peak in the total signal amplitude was observed at about 6 K, but cooling in a field of 3 kG to $2 \cdot 6$ K and subsequent warming up betrayed no differences from a zero-field-cooled alloy.

Lower concentrations ($0 \cdot 02$, $0 \cdot 11$ and $0 \cdot 49 \%$ Mn in Mg) were studied at both X-band and K-band by Kleinhams and Wigen (A 88). By studying the line broadening as a function of temperature for the different concentrations the system was shown to be bottlenecked. In the $0 \cdot 02 \%$ Mn in Mg alloy a Korringa slope of about $40 \, \text{G.K}^{-1}$ was obtained, reducing to approximately $5 \, \text{G.K.}^{-1}$ for the highest concentration. These results put a lower bound on the value of J_{eff} for the system of $0 \cdot 12$ eV. At low temperatures in all the alloys line broadening associated with magnetic ordering was observed and the resonance was shifted rapidly to lower fields; the observed frequency dependence of this shift, however, was rather less than the inverse dependence on frequency predicted by the equations for an antiferromagnetic transition of Keffer and Kittel (1952).

The Mn in Zn system, from an E.S.R. point of view, still remains very much an enigma. Collings *et al.* (A 31) claim to have observed a resonance from a powdered $0 \cdot 4$ at. % Mn alloy with a linewidth $\sim 800 \pm 10$ G at $4 \cdot 2$ K, a shallow minimum in the linewidth occurring at roughly 10 K, which was taken as indicative of antiferromagnetic ordering. Miyako (A 97) also claims to have observed resonance from powder samples of $0 \cdot 5$ and $1 \cdot 0$ at. % Mn. In this case a g-value of $2 \cdot 05 \pm 0 \cdot 01$ was observed, again with a linewidth of 800 G at $4 \cdot 2$ K which was then reported to broaden at higher temperatures. Addition of Al and Cu impurities to the $1 \cdot 0$ at. % alloy led to a larger residual width and an increased temperature dependence of the linewidth. Since, however, the solubility of Mn in Zn is reported to be no more than $0 \cdot 4$ at. %, this latter work should perhaps be treated with some reservations. There are a host of dilute intermetallic compounds in the Zn–Mn system (see below) and it is possible (particularly bearing in mind that resonance was observed by Miyako up to 90 K) that one of these was in fact stabilized rather than the solid solution. The

present author and Orbach's group (private communication) have been unable
to detect a resonance of Mn in Zn for powder samples.

Devine and Moret (A 61), however, do claim to have observed a resonance
from a single crystal of Zn containing 296 p.p.m. of Mn. A g-factor for the
external magnetic field parallel to the c-axis of $2 \cdot 0065 \pm 0 \cdot 005$ and for the field
perpendicular to the c-axis of $2 \cdot 0015 \pm 0 \cdot 0005$ was obtained. Over the range
of their measurements ($1 \cdot 5$ K to 26 K) these latter values, corresponding to
higher temperatures, were found to decrease to $1 \cdot 998$ and $1 \cdot 993$, respectively as
the temperature was lowered. They suggest that this decrease in g-value, like
the increase in the exceedingly narrow linewidth from 10 G at 26 K to 13 G at
2 K, was due to the onset of magnetic ordering. Whilst the increase in linewidth
may possibly be attributed to such effects, it would appear that the g-shift is
moving away from the direction normally associated with ordering effects.

4.11. *Intermetallic compounds containing Mn*

In contrast to the wealth of experimental data which exist for rare-earth
s-state ions in intermetallic compounds, very few results exist for Mn in intermetallic compounds.

Results on the few compounds of this sort which have been studied have
proved exceedingly interesting. The compound $Zn_{13}Mn$ was shown to have
rather remarkable E.S.R. properties (A 20, A 21, A 151). At high temperatures
(> 120 K) the compound has a g-value close to the Mn free-ion value but on
lowering the temperature a progressive increase occurs which takes the g-value
to a value close to 3 in the liquid helium temperature range (see fig. 33). Apart
from the wide range of temperature over which the shift occurs, there would be
nothing remarkable about this result if it were not for the fact that magnetization
measurements indicate that the shift cannot be accounted for by the effects of a
demagnetizing field. The Mn ion in this compound only has a small ($\sim 0 \cdot 3 \mu_B/$Mn
in 3 kG) moment and such effects could account at most for only 20% of the
observed shift. The results seem to reflect the limited dimensionality of this
compound. At high temperatures it is believed that alignment of Mn atoms
takes place along chains in one direction and that at a lower temperature
(~ 22 K) the chains lock together to form a true three-dimensional ordered
structure. Substitution of Mn atoms on Fe sites in a sample of $Zn_{13}Fe$ (thus
destroying the ionic uniformity of the chains) results in a loss of the resonance.

In the study of the E.S.R. behaviour of magnetic ordering in compounds,
Taylor and Coles (A 148, A 151) reported results on two further compounds.
The Heusler alloy Pd_2MnSn which orders ferromagnetically at 189 ± 3 K showed
E.S.R. behaviour similar in form to that of the compound $GdAl_2$ which orders
at a temperature of ~ 168 K; the line began to broaden and shift at ~ 220 K.
A g-value of $2 \cdot 013 \pm 0 \cdot 005$ was obtained above this temperature and in the same
temperature regime the linewidth broadening at a rate of $4 \cdot 5$ G.K^{-1}.

One interesting feature of this result was that the high temperature linewidth
extrapolates unequivocally to a large negative residual linewidth. The δ-phase
compound (Dunlop and Taylor, to be published) of Mn in Zn, roughly $Zn_{10}Mn$,
shows somewhat similar behaviour to the Heusler alloy.

Fig. 33

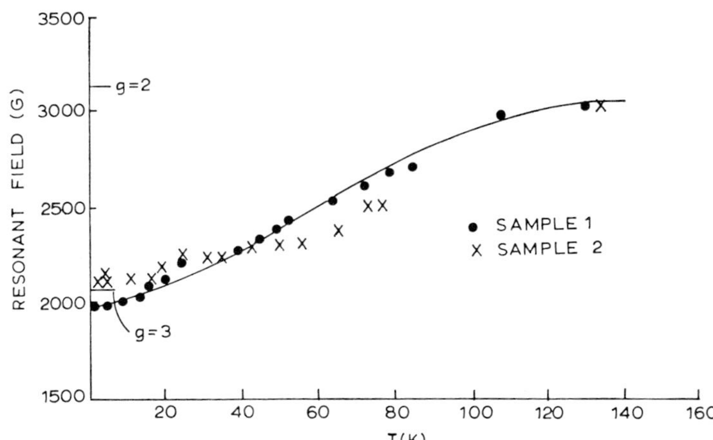

The g-shift as a function of temperature for the compound $Zn_{13}Mn$ (after Caplin *et al.*, A 20).

Devine *et al.* (A 66) studied E.S.R. of Mn dissolved in concentrations of 400 p.p.m. to 4000 p.p.m. in the compound $MoGa_4$. The g-value and linewidth showed an extremely strong temperature dependence, the form of which did not change with Mn concentration. The g-value was approximately temperature independent at $g = 2$ above 30 K but then increased steeply to $g \sim 2\cdot15$, whilst the linewidth increased from about 100 G at 80–100 K to nearly 1 kG at 4·2 K. The authors attempted to analyse their data in terms of a p-virtual bound state localized to the ion and argued that at high temperatures there exists a bottleneck between the composite ion and the conduction electrons which they suggest is broken at low temperatures by a Kondo effect. The authors rule out the possibility of the observed effects being due to magnetic ordering but the concentration independence and the form of the data as a function of temperature strongly suggest the possibility of the separation of small quantities of a second phase (perhaps $MnGa_4$).

4.12. *Other 3d elements as impurity*

In § 2 we discussed some of the possible reasons why E.S.R. in metals has not been extended into the study of the other 3d elements in various hosts. There have from time to time been reports of resonances having been observed from, for example, Co in (R.E.) Co_2 compounds (Barnes *et al.*, 1966) and from Fe in dilute Cr–Fe alloys (A 128). The former result has been shown to be spurious. There have been a number of cases such as Al–Fe and Mg–Fe (A 30) where resonances have been reported as absent. One alloy in which an Fe local moment resonance might have been detected is Fe in the compound $Ce\,Ru_2$ (A 91). In this case Koopmann *et al.* observed a resonance with a g-value of $2\cdot65 \pm 0\cdot05$ for samples containing 250 to 5000 p.p.m. of Fe. This g-value appeared to be (see figs. 34, 35) temperature independent when followed to 10 K in the higher concentration alloys. The linewidth showed a Korringa slope $\sim 40\,\text{G.K}^{-1}$. The authors stated that the signal was much stronger in the superconducting state than the normal state, with a change of A/B ratio from 2·2 to 1·2 at the transition.

Fig. 34

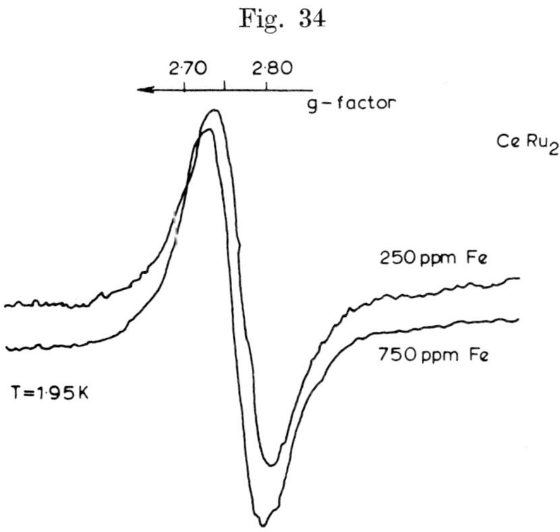

Typical spectra of Fe in $CeRu_2$ (after Koopmann *et al.*, A 91).

Fig. 35

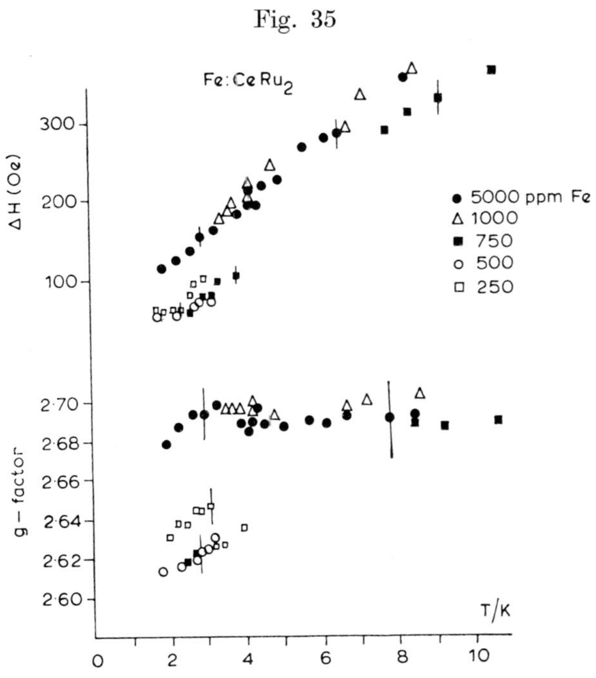

g-factor and linewidth as a function of temperature for Fe in $CeRu_2$ (Koopmann *et al.*, A 91).

If the reason for the general absence of an observable resonance from 3d elements in metals apart from Mn is due to the fact that they are unbottlenecked (Hirst 1972) we might anticipate that the best chance of observing (say) an Fe resonance would be at low temperature in an exchange-enhanced host where the effects of enhancement are to pull down the unbottlenecked value of the Korringa slope. It is probably this mechanism that allows us to detect a resonance of Mn in Pd at working temperatures. It is possible that hosts like $CeRu_2$ may satisfy such a criterion.

Resonance has, of course, been observed for Cr in Cu by the transmission technique but there have been no reports of a resonance from Cr in Cu using reflection E.S.R.

There have been a number of observations of E.S.R. of Co, Fe and Ni above the Curie temperature in ferromagnetically ordering alloys like those, for example, of Co in Pt and Pd, Fe in Pd, Ni in Cu and W by Bagguley and Heath (1967) and Bagguley *et al.* (1967), but these measurements do not fall within the scope of the present article. Resonance of Fe in V and Ti alloys, which are not ferromagnetic but in which the Fe atoms form superparamagnetic clusters, have been reported by Burr *et al.* (A 11) and recently (to be published) Taylor and Murani have studied E.S.R. of superparamagnetic clusters of Fe and Au. Burr and his co-workers only observed resonance from alloys of Fe in V with concentrations in excess of 24 at. % ; a 26% alloy gave a high temperature g-value of $2 \cdot 18 \pm 0 \cdot 01$ which remained constant down to 40 K but decreased to $2 \cdot 14$ at lower temperatures. The linewidth in the same alloy broadened from 800 G at 140 K to ~ 1400 G at 4·2 K and the linewidth was found to be strongly frequency (field) dependent, more than doubling when the frequency was increased from 35 GHz to 70 GHz. This was ascribed to inhomogeneous broadening.

Apart from the study of these concentrated alloys, where E.S.R. may make a valuable contribution to our understanding of the state of the impurities, it would seem that our knowledge of Fe, Cr, Co and Ni as dilute solutions in various hosts is likely to be advanced from E.S.R. measurements chiefly by experiments like the double bottleneck work of Hirst *et al.* (A 84) on Cr in Cu Mn. The attainment of really low temperatures for E.S.R. experiments may, of course, increase the scope of direct measurements on alloys of this kind.

4.13. *Non S-state impurities*

In this section we rather arbitrarily put together the results on Dy, Er, Yb and the one reported Ce resonance in various hosts. We have not attempted to split this section into subheadings for each of the impurities, because in general the studies of these non s-state ions have been carried out as a group in a particular host metal.

The first observation of a non S-state paramagnetic resonance in a dilute alloy was made in an Ag 0·3 at. % Er alloy by Griffiths and Coles (A 79) at X-band. A resonance line with $g = 6 \cdot 73 \pm 0 \cdot 1$ was obtained at 2·5 K, close to the value observed for trivalent Er in Cubic non-metallic matrices, where the $^4I_{15/2}$ state is split by the crystal field to give a Γ_7 ground-state doublet. The linewidth was observed to broaden from about 60 G at 2 K, linearly to 160 G at 14 K. Over this temperature range the signal intensity $(A + B)\Delta H^2$ showed little temperature dependence. A 1·5 at. % sample showed weaker temperature dependence of the linewidth but gave the same g-value. In a further paper (A 83) an improved g-value of $6·80 \pm 0·07$ was reported for the Er alloy. A resonance was also reported for 0·9 at. % Yb in Au with a g-value of $3·30 \pm 0·10$, close to the value obtained for trivalent Yb in a cubic dielectric host (3.43). The linewidths of both alloys were found to be well fitted to the form $A + BT$ with $A = 42$ G and $B = 9·2$ G . K^{-1} for the Ag–Er alloy and $A = 370$ G, $B = 12$ G . K^{-1} for the Au–Yb. No resonances were observed in powder samples of Ag Dy, Ag Yb or from Tm, Ho and Tb. The absence of resonance from the Ag–Yb was attributed to the divalent character of Yb in this alloy, as was also suggested by the small and weakly temperature dependent susceptibility.

A resonance from Er in a single crystal of Mg was reported by Burr and Orbach (A 10). The observed signal was characteristic of a uniaxially distorted Γ_7 doublet of the free ion $^4I_{15/2}$ multiplet, the principal values of the g-tensor being $g_{\parallel} = 8·90 \pm 0·01$ and $g_{\perp} = 5·577 \pm 0·01$. The broadening of the linewidth as a function of temperature was observed to be anisotropic, varying from 120 ± 20 G in the parallel direction to 140 ± 20 G in the perpendicular direction at 4·2 K. Analysis of these results was taken to suggest that the conduction electron–localized moment exchange interaction was isotropic with a value of 0·031 eV, close to the value (0·03 eV) obtained by Griffiths and his co-workers for Ag–Er.

The first observation of hyperfine splitting of a localized moment in a metal was made by Chui *et al.* (A 26) for concentrations of from 65 to 1500 p.p.m. Er in Ag. A value of $A^1 = 75 \pm 10$ G for the hyperfine coupling constant was obtained, in good agreement with values for Er in insulator hosts. The authors observed that the residual linewidth (A) was concentration dependent for these alloys, and attributed the failure of previous measurements to detect hyperfine splitting as due to the larger residual width for the higher concentrations used. In a further paper (A 146), hyperfine splitting was also observed in dilute Au–Er (to 50 p.p.m. Er) and in Au–Yb (to 500 p.p.m. Yb). A value of $75·5 \pm 0·5$ G was obtained for Au–Er (cf. the value of 73·8 for Er^{3+} in ThO$_2$), the central line gave $g = 6·80 \pm 0·04$ and the linewidth obeyed the relation $A + BT$ with $B = 2·4 \pm 0·2$ G . K^{-1}. The hyperfine structure in Au–Yb corresponded to the splitting of the $I = \frac{1}{2}$, ^{171}Yb isotope and in this case a value $A^1 = 575 \pm 10$ G (cf. $A^1 \sim 549$ G for Yb in CeO$_2$) was obtained. The hyperfine splitting from the isotope ^{173}Yb, which has about the same natural abundance as the ^{171}Yb isotope was not observed and this was attributed to the reduction in intensity on division between the six satellites of the multiplet. A g-value of $3·34 \pm 0·06$ was found for the central line, but the value of B, the temperature-dependent contribution to the linewidth, was found to be 42 ± 7 G . K^{-1}—a factor of about three greater than the value reported by Hirst *et al.* (A 83) for the slightly more concentrated alloy. The increase in A^1 for both alloys over the values in insulators was attributed to the spin polarization of the conduction electrons. The localized moment–conduction electron exchange was deduced from the thermal broadening of the linewidth and using a value estimated by Gossard *et al.* (1962) for the 6s contribution to the hyperfine coupling constant in Yb metal, contributions were separated out for the atomic and covalent mixing contributions to J for both alloys. By this method a value for the covalent mixing in Au–Yb some 20 times larger than theoretically predicted was obtained and the authors suggested that (bearing in mind particularly the diamagnetism of Ag–Yb) this difference in the contribution in the covalent mixing for Au–Yb is due to the close proximity of the 4f level to the Fermi level in this alloy.   The values for A^1 given in the above paper were subsequently revised (A 25) when samples made with enriched ^{167}Er and ^{171}Yb became available.   The revised values of A^1 were 75 ± 0.5 and 575 ± 10 and for B, 2.7 ± 0.5 and 40 ± 10 G . K^{-1} respectively. Clearly these latter values still support the above analysis.

The resonance of Dy in Ag was observed by Davidov et $al.$ (A 43) for Ag concentration of from 0.05% to 1% at temperatures in the pumped He^4 range. A g-value of 7.80 ± 0.3 (close to the Γ_7 doublet value of 7.55 for Dy in CaF$_2$) was obtained and the linewidth followed the usual linear relationship as a function of temperature with $A(C) = 230$ G for a 0.4% alloy and 130 G for a 0.1% alloy. The value of the thermal broadening was constant at 40 ± 15 G . K^{-1}.   No hyperfine structure was observed.

The first observation of a Γ_8 quartet resonance was reported by Davidov et $al.$ (A 41) in Au–Dy single crystals with concentrations of 500 and 1000 p.p.m. Dy.   The 500 p.p.m. sample displayed two lines, both exhibiting appreciable angular dependence in resonant field.   The 1000 p.p.m. sample showed only one line with angular variation similar to that of one of the 500 p.p.m. sample lines. The data were fitted to crystal-field parameters with $B_4/B_6 = 1300 \pm 30$.   This value is much larger than the value $B_4/B_6 \leqslant 340$ (deduced from the work of Hirst and Williams) for an AgDy alloy and the difference was attributed to the larger virtual d-level width in Ag, reducing the contribution to B_4 from the d-like screening electrons.   The nearly linear increase with temperature of the line-width of the strongest line in the most dilute crystal was used to evaluate $J = 0.14 \pm 0.05$ eV.   This value (much larger than Au–Er) was taken to be evidence, along with the measurements described above, of the increasing tendency towards covalent mixing in moving across the heavy rare earths from Gd towards Yb.

The UCLA group (A 53) also investigated the resonance of Er as impurity in Th, Rh, Pt and Cu and their results are summarized below:

	C(p.p.m.)	g	A	B
Th–Er	100	6.84 ± 0.05	27	15 ± 2
Rh–Er	700	6.78 ± 0.1	63	4 ± 2
Pt–Er	1000	6.50 ± 0.03	500	--
Cu–Er	200	6.86 ± 0.07	54	10 ± 2

Hyperfine structure was observed in the Th–Er system but (presumably due to the larger residual linewidth) was not observed in the other alloys.

These measurements were extended in a further paper by the same authors (A 48) where the behaviour of Er in a number of exchange-enhanced hosts was analysed.   Powder samples of Er in Pt gave broad lines with $g = 6.5 \pm 0.3$ as found previously, but splat-cooled samples gave a narrow line with $g = 5.95 \pm 0.05$. Hyperfine structure was observed with $A^1 = 75 \pm 2$ G.   All concentrations from 250 to 1020 p.p.m. gave a linewidth linear in temperature with $B \sim 8$ G . K^{-1}. Pd–Er samples were found to give two lines in powder samples with Er concentrations from 200 p.p.m. to 10 000 p.p.m.   The g-values of these two lines were quoted as 7.5 ± 0.05 and 4.5 ± 0.5.   It was found that splat-cooling the samples reduced the intensity of the first line and increased that of the second (low g), and hydrogenation of the alloys caused the second line to vanish. From these observations the broad line in the Pt–Er powder samples and the $g = 7.5$ line in the Pd–Er samples were attributed to intermetallic compounds. It was argued that although the Γ_6 level has $g = 5.95$, the associated value of the hyperfine coupling constant A^1 would be 12% lower than the observed value. Since Er in seven other hosts investigated up to this time all gave a value of A^1 less than 1·5% from the expected value, it was argued that the Er resonance observed in Pd must be due to the Γ_7 level but shifted from the free-ion value of 6·78 by some 34%! A plot of Gd and Er g-values against host (Rh, Pt, Pd) susceptibility yielded a pair of straight lines. The values of the g-shifts, however, do not predict through the Korringa relation, the observed temperature dependence of the linewidth even allowing for exchange enhancement.

This problem of non-S-state rare earths in Pd has been investigated in detail by the Geneva group. Devine *et al.* (A 67) studied the E.S.R. spectrum of single crystals of 400, 800 and 1600 p.p.m. of Er in Pd. Hyperfine splitting was observed, leading to a value $A^1 = 64 \pm 5\,\text{G}$. The angular variation of the field for resonance of the most intense line and for the various transitions of a quartet agreed well with the observed form of the lines, suggesting that the ground state of Er in Pd is neither the Γ_7 nor Γ_6 doublets as considered by the UCLA group in their powder measurements, but is rather one of the three possible Γ_8 quartets. From measurements of the g-shift and slope of the linewidth, an exchange constant J of $-0.014 \pm 0.002\,\text{eV}$ was deduced, in line with the values obtained for Gd in Pd. In a recent re-examination of the Pd–Er system, using crystals grown by the floating-zone technique rather than by recrystallization, Zingg

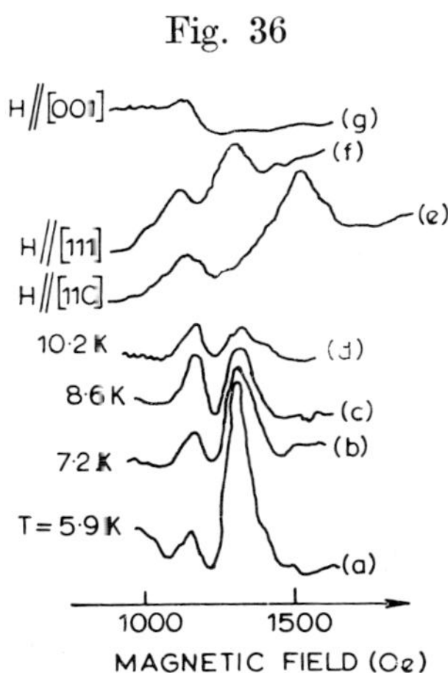

Fig. 36

E.S.R. spectra of Er^{3+} in Pd. Spectra *a* to *d*, obtained with sample *A* (1500 p.p.m.) and **H** along (111). The low field line corresponds to the excited doublet (transition 5→6) and the high-field one to transition 2→3 of the $\Gamma_8^{(3)}$ ground quartet. The increase of the relative intensity of the excited state with temperature is readily seen. Spectra *e* to *g* obtained with sample B (1000 p.p.m.) at constant temperature $T = 9.3$ K. The high-field line (transition 2→3) changes position according to the orientation of H whereas the low-field line (transition 5→6) has an isotropic g-value). (Zingg *et al.*, A 160.)

and his co-workers at Geneva (A 160) have observed a new isotropic line (see fig. 36) with $g = 5\cdot8 \pm 0\cdot1$ which they have identified as being due to the first excited Γ_6 doublet. The amplitude of the signal from this line increased strongly relative to the Γ_8 quartet resonance and comparison of the intensities has allowed the determination of the energy scaling parameter, W, of the crystalline electric field. A more precise value $(-0\cdot163 \pm 0\cdot015)$ was deduced from the splittings between the levels within the Γ_8 quartet as a function of orientation. The crystal-field parameters in this work were found to have signs opposite (both negative) to the predictions of the point-charge model (and in the case of the sixth-order parameter, to the virtual bound-state model as well).

The resonance behaviour of Dy in Pd has raised considerable controversy. Devine *et al.* (A 63) first reported observation of the Dy resonance in single crystals of Pd grown by the recrystallization technique. The spectrum contained four main lines, one of which was strongly anisotropic and this led the authors to ascribe the lines to a Γ_8 quartet (as Pd–Er). Weak hyperfine lines originating from ^{161}Dy and ^{163}Dy isotopes were observed but their low intensity did not enable values of the hyperfine constants to be extracted. The linewidth of the most intense line in the (110) direction was found to be linear as a function of temperature, with a residual width of 10 ± 1 G and a slope of $45\cdot0 \pm 0\cdot5$ G . K^{-1} for both 100 p.p.m. and 500 p.p.m. samples.

Although a good fit could be made to the angular dependence of the most anisotropic line in the spectrum by using Bleaney's Hamiltonian for the quartet, an equally good fit could not be obtained for the other lines; although the general form of the lines as a function of orientation was given fairly well, the magnitudes were not. The authors rejected the possibility that the next excited level (another Γ_8) were admixed into the ground state and that the bad fit might be due to 'hopping' effects. Instead they suggested that the insulator Hamiltonian is inadequate for this system in not including the effects of local moment–conduction electron exchange.

The question of this poor fit was taken up by Praddaude (A 119). He suggested that a fit to the central line using the full crystal-field Hamiltonian could be obtained but with values of W and X which imply closely spaced energy levels and thus that the spin Hamiltonian approximation will not apply in this system. An estimated splitting between the $\Gamma_8^{(3)}$ and $\Gamma_8^{(2)}$ levels of $2\cdot5$ K was deduced, thus implying that mixing effects from the applied magnetic field will be strong. The calculation further suggested that the most intense transition occurred between the second and third levels of the $\Gamma_8^{(3)}$ quartet and not between the first and second as assumed by Devine *et al.* In the case of Pd–Dy the values of W and X obtained from E.S.R. appear to be in strong disagreement with the values obtained by fitting crystal-field theory to the high field magnetization data for the system (A 120). Comparison of measurement of magnetization and E.S.R. on Pd-based alloys appear to be fraught with contradictions (see Pd–Gd above).

In a further paper, Devine (A 62) re-interprets the earlier data for the $\langle\frac{1}{2}\rangle\rightarrow\langle-\frac{1}{2}\rangle$ quartet transition but rejects the argument of Praddaude that admixture effects are important. He shows that inclusion of conduction electron–local moment exchange to the Hamiltonian (as suggested but not included in the previous paper (A 63)) leads to a value of $B_4/B_6 = 292 \pm 4$. In this re-analysis, a second resonance reported in Devine's earlier paper to be isotropic is shown in fact to be slightly anisotropic.   It was suggested that this line originates from Dy ions in sites of non-cubic symmetry possibly as some form of interstitial. It was further suggested that the perturbation of the cubic crystal field caused by these ions might be sufficient to broaden beyond observation the other lines from the quartet which are missing.

Analysis of the behaviour of rare earths in other elemental hosts such as Th, Al, Rh and Ir has been studied in detail by the UCLA group.   Th–Er alloys containing from 40 to 2600 p.p.m. yielded a line well fitted by the formula $\Delta H = A(c) + BT$, with a value of $B = 14 \pm 3$ G.K^{-1} (A 40).   A g-value of $6 \cdot 84 \pm 0 \cdot 05$ (as previously reported for a single 100 p.p.m. sample) was obtained and hyperfine structure was resolved with $A^1 = 75 \pm 2$ G.   In Th–Dy alloys with concentrations in the range 180 to 3250 p.p.m. Dy, values of $g = 7 \cdot 61 \pm 0 \cdot 1$ and $B = 33 \pm 7$ G.K^{-1} were reported independent of concentration.   The temperature dependence of the linewidth in these alloys was used to estimate J with an assumed enhancement factor, $K(\alpha) \sim 0 \cdot 4$.   The values obtained were within the errors of the values obtained from the g-shift (with the exception of Th–Gd as reported in §4.2) and yielded a constant value $\sim 0 \cdot 044$ eV for all three alloy systems, in good agreement with the values obtained from previous superconducting transition temperature versus rare-earth concentration measurements.   Alloys of up to 1000 p.p.m. of Dy and Er in Al (as well as Gd) yielded g-values of (A 122) $7 \cdot 58 \pm 0 \cdot 05$ and $6 \cdot 82 \pm 0 \cdot 04$, respectively, both values in agreement with those expected for a Γ_7 doublet ground state.   Hyperfine structure (fig. 37) was observed in both cases giving values of A^1 of $75 \cdot 5 \pm 1$ G for ^{167}Er and 82 ± 4 G for ^{163}Dy.   The slopes and residual linewidths were approximately concentration

Fig. 37

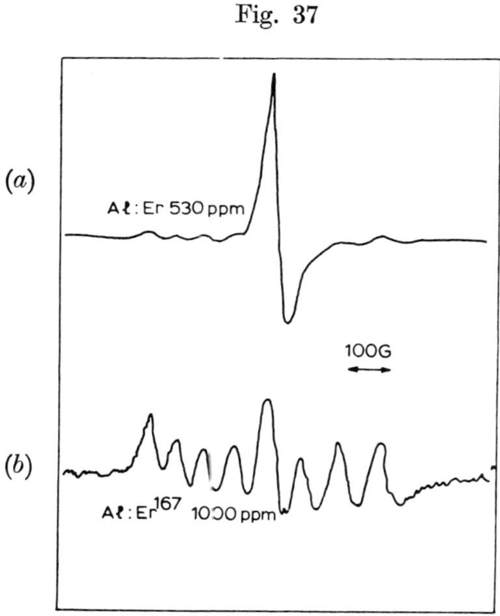

(*a*) Electron spin resonance spectrum of Er in Al, for a natural abundance of Er at a nominal concentration of 530 p.p.m. at $1 \cdot 4$ K.   (*b*) Spectrum for Er^{167} in Al for a concentration of 1000 p.p.m. at $1 \cdot 4$ K.   The hyperfine spectrum is now clearly resolved (from Rettori *et al.*, A 122).

Fig. 38

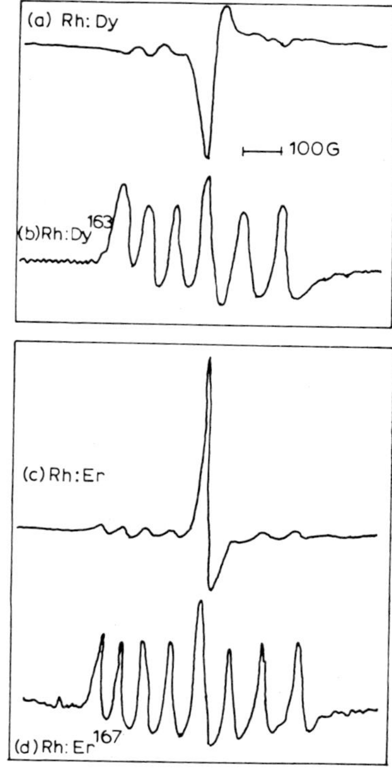

(a) The electron spin resonance spectrum of RhDy for a natural abundance Dy (nominal concentration of 500 p.p.m. at 1·4 K). (b) The E.S.R. spectrum of Rh–^{163}Dy for a ^{163}Rh concentration of 1000 p.p.m. at 1·4 K (after Davidov *et al.*, A 53).

independent with $A = 40$ G and $B = 28$ G.K^{-1} for Al–Dy. The results were the same for both arc-melted and splat-cooled samples. In order to estimate the contribution to A from unresolved hyperfine structure originating from other Dy isotopes (e.g. ^{161}Dy with a natural abundance of 18·9%), alloys containing ^{164}Dy with $I = 0$ were measured—these yielded $A = 33$ G, showing that a small part of the residual width in Al–Dy, using the naturally abundant distribution of Dy isotopes, derived from this source (i.e. up to 7 G). No influence of higher excited levels of the crystalline electric field was detected in either system and a value of $J = +0·11 \pm 0·02$ eV was extracted from both Δg and B, for both types of alloy. It was in fact in the Rh–Dy alloy system that Davidov *et al.* (A 53 and fig. 38) detected the first hyperfine splitting of ^{163}Dy, the results for the Er and Dy in Rh systems for alloys containing to 1000 p.p.m. are shown for two typical alloys below:

	C (p.p.m.)	g-value	A	B	A
Rh–^{163}Dy	500	$7·51 \pm 0·1$	15	8 ± 3	$85 \pm 0·5$
Rh–^{167}Er	200	$6·80 \pm 0·05$	8	$4·5 \pm 3$	$75·5 \pm 0·5$

These values of hyperfine constant are a little larger than for the same isotopes in non-metals. The two alloys for which the results are shown above were splat-cooled in preparation and in such cases the residual linewidth were found to be exceedingly small. It was in this paper that the authors argued that the sign of the contact hyperfine field at the nucleus disproves the d–s hybridization mode of Coles *et al.* (A 28) for the sign of the exchange integral. As discussed in § 2.1, however, this argument seems to neglect the contribution of core polarization effects. Powdered and splat-cooled samples of Er and Dy in Ir (A 50) gave extremely narrow lines (although curiously only a very broad line was observed for Gd in Ir). The ground state of both the Er and Dy were ascribed to the Γ_7 doublet and alloys in the concentration range 150 to 2000 p.p.m. gave g-values of $6 \cdot 74 + 0 \cdot 05$ and $7 \cdot 49 \pm 0 \cdot 07$, respectively, with average values of A of $3 \cdot 2 \pm 1$ in Ir–Er and $7 \cdot 0 \pm 2 \cdot 0$ for Ir–Dy. Hyperfine constants of $75\,G$ for ^{167}Er and $82\,G$ for ^{163}Dy were reported.

The resonance of non-S-state rare earths in hosts with band structures analogous to the rare earths (such as La, Y, Sc, Lu) but which are not themselves magnetic, have received little attention. Alekseevskii *et al.* (A 1) studied Er concentrations of from $0 \cdot 5$ to 3 at. % in superconducting La. The g-value was quoted as $6 \cdot 83 \pm 0 \cdot 05$; the resonance resulting from Er in the β–La (cubic) host phase rather than the α-hexagonal host phase. On passing through the superconducting transition the g-factor was not found to change but the linewidth decreased sharply at T_c, then increasing with decreasing temperature at lower temperatures. The slopes of the linewidth as a function of concentration (up to 3 at. %) was shown to be greater at $4 \cdot 2\,K$ for the normal phase than for the superconducting phase at $2\,K$ and $3\,K$. The only other measurement on alloy systems of this type has been recently performed by Keller *et al.* (A 86). Magnetization measurements on single crystals of Er in Sc enabled calculation of the crystalline electric-field parameters and energy levels. E.S.R. was used to confirm that the Er ground state was a Γ_7 doublet (the point-charge model would predict a Γ_9 ground state) and the sample containing 2000 p.p.m. Er gave values of $g = 10 \cdot 4 \pm 0 \cdot 3$ for the field parallel to the crystal c-axis and $g = 5 \cdot 02 \pm 0 \cdot 06$ for the field along the a-axis. Good agreement was obtained between the observed angular dependence of the E.S.R. resonant field and the predicted form based on the magnetization measurements.

Recently work has begun on the analysis of the crystal-field splittings of rare-earth ions in the simpler forms of intermetallic compound. Devine *et al.* (A 64) studied the spectra of Er and Dy in YAl_2 and observed in addition a resonance from an $(Y_{95}Tm_5)\,Al_2$ alloy which they believed might be due to the $\Gamma_5^{(1)}$ triplet level of Tm. This latter interpretation, however, was made difficult to sustain by virtue of the fact that no temperature regime was found in which the linewidth was linear (indeed it broadened continuously on decreasing the temperature from $20\,K$ to $4\,K$) and a temperature-dependent g-value was observed approaching $2 \cdot 22 \pm 0 \cdot 02$ in the high temperature limit compared with the calculated value of $3 \cdot 36$ for the Γ_5 state. A single crystal of YAl_2 containing $0 \cdot 3$% Er showed one weak line at temperatures below $2\,K$. A g-value of $7 \cdot 1 \pm 0 \cdot 1$ was estimated and the authors attempted a crude estimate of the slope of the linewidth over this very limited temperature regime, obtaining a figure of $183 \pm 84\,G.K^{-1}$. In a powder sample, a line with $g = 6 \cdot 8 \pm 0 \cdot 2$ and an approximately temperature independent linewidth of $500\,G$ was observed. The form of the linewidth in this case, it was suggested, might be due to either strains or second phase. A possible large positive g-shift was also observed for a Dy sample; here the doublet state gave $g = 8\cdot0 \pm 0\cdot5$ ($\Delta g = 0\cdot5 \pm 0\cdot5$) and the linewidth broadened with a slope of 51 G.K^{-1}. The E.S.R. results, it was pointed out by the authors, in ascribing Γ_7 ground states to both Er and Dy in YAl$_2$ appear to contradict the neutron-diffraction results on the system (Purwins et al. 1966) which suggest a Γ_5 triplet ground state. It will clearly be of considerable interest to analyse further the line from the Tm substituted compound; with full analysis it may be that the Γ_5 ground state ascribed to Tm in YAl$_2$ in the neutron measurements is also wrong.

Davidov and his co-workers (A 49) studied the E.S.R. behaviour of Ce^{3+}, Dy^{3+}, Er^{3+} and Yb^{3+} in various cubic intermetallic compounds. A ground-state resonance was observed for Yb in a LaPd$_3$ powder sample with $g = 3\cdot34 \pm 0\cdot01$ and a hyperfine constant for the ^{171}Yb nucleus of $A^1 = 570 \pm 15$ G. The small negative g-shift observed here would appear to be in contrast both with the enhancement of A^1 over the value for an insulator host and with the small positive g-shift shown by Gd in LaPd$_3$ ($g = 2\cdot01$). Single-crystal measurements of the compounds LaSb, LaBi, LuSb and LuBi containing 2000 p.p.m. and 500 p.p.m. of Er gave highly anisotropic lines which were fitted by assuming a Γ_8 quartet ground state. Hyperfine structure associated with ^{167}Er ($I = 7/2$) gave $A^1 = 75\cdot5 \pm 1$ G. It was in LaSb that Davidov was able to report the first observation of a Ce^{3+} resonance in a metallic host. A Γ_7 ground-state doublet gave a g-value of $1\cdot43 \pm 0\cdot02$, which is close to the theoretically predicted value of $1\cdot42$. Both Dy and Yb in LaSb were shown to have Γ_6 doublet ground states with $g = 6\cdot70 \pm 0\cdot1$ and $2\cdot50 \pm 0\cdot02$, respectively. These observations are consistent with positive values of the fourth and sixth-order crystal-field parameters.

Er as impurity in LaB$_6$ was found to give rise to a single line with $g = 5\cdot90$; close to the value of $5\cdot85$ expected for a Γ_6 doublet ground state. The large Korringa slope of 14 ± 5 G.K^{-1} was attributed partly to the effects of the low-lying Γ_8 level. According to the tabulations of Lea et al. (1962), the fact that a Γ_6 ground state is observed would suggest that either the sixth-order term and the fourth-order crystal-field terms are negative or that the fourth-order term is negative and sixth order positive. These results and the others, above, were discussed in detail in terms of the various crystal-field models. In the rocksalt structure La compounds, good agreement with the point charge-model is observed, whilst for Er in LaB$_6$ the ambiguity in signs of $A_4\langle r_4 \rangle$ and $A_6\langle r_6 \rangle$ makes it impossible to distinguish between the possible validity of a simple point-charge model or of the importance of the 5d virtual bound state.

We conclude this section by returning full circle to the results of two recent experiments on non-S-state rare earths in 'simple' metal hosts. Firstly, that of Davidov and the group at UCLA on the effects of low-lying crystalline-field excited states on the E.S.R. linewidth (A 57) (see also Baberschke, proceedings of the Haute-Nendaz Conference (A 6)). Powder samples of Au-Er were found to show a sharp change (see fig. 39) in slope in their linewidths at about 6 K and this was interpreted as being due to the proximity of the $\Gamma_8^{(1)}$ level to the Γ_7 ground state and enabled their separation to be estimated. These measurements, together with the direct observation of the resonance from an excited state reported recently in Pd Er by Zingg et al. (see above), constitute the only cases where E.S.R. has been able to provide this important information.   A single crystal of AgDy was also measured which gave $g = 7.55$ and yielded a hyperfine constant $A^1 = 82 \pm 4$ G for ^{163}Dy, but in this case no change in slope of the linewidth was observed.   Williams and Hirst (1969) obtained from their susceptibility measurements a separation between the Γ_7 and $\Gamma_8^{(1)}$ levels of less than 1 K for AgDy but the authors argue that separations smaller than 5 K would lead to observable anisotropy in the E.S.R. g-value.   An upper limit was, on the other hand, perhaps rather more questionably obtained by comparing the temperature dependence of the thermal broadening of Dy and Er in other cubic hosts, where except for the hosts Al and Au, $B(\mathrm{Dy})/B(\mathrm{Er}) \sim 2.5$.   The value of 3·0 for Ag as a host, it was argued, would place an upper limit of 10 K on the Γ_7–Γ_8 splitting.   In Au : Er the change in slope of the linewidth at 6 K leads directly to a value of 16 ± 6 K for the separation of the excited $\Gamma_8^{(1)}$ state from the Γ_7 ground-state doublet.   This result is in excellent agreement with the magnetic susceptibility results of Williams and Hirst.

Fig. 39

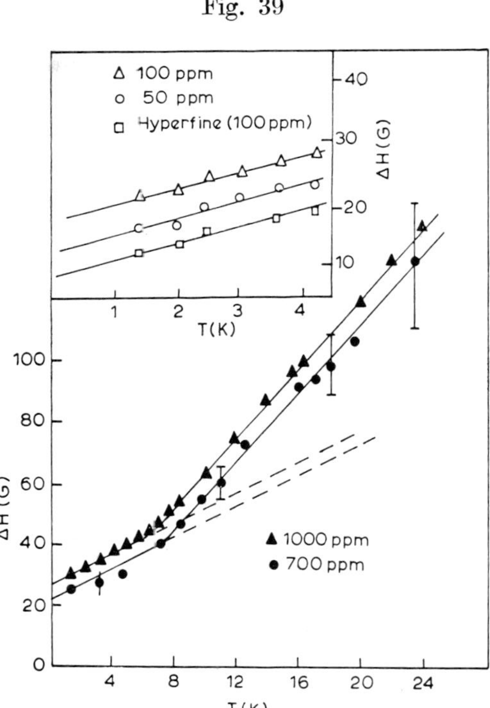

The E.S.R. linewidth as a function of temperature for several Au : Er alloys.   The solid line represents the theoretical linewidth (modified to include the experimental residual width) for $\Delta = 16$ K.   The dashed line is the linewidth expected for an isolated Γ_7 doublet.   The insert exhibits the linewidth at low concentrations in the temperature range $1.4 \leqslant T \leqslant 4.2$ K.   The slope of the thermal broadening agrees in this temperature range with the dashed line, i.e. is indistinguishable from an isolated Γ_7.   The squares in the insert represent the linewidth extracted from one of the hyperfine lines of the Au : Er sample (Davidov *et al.*, A 57).

Fig. 40

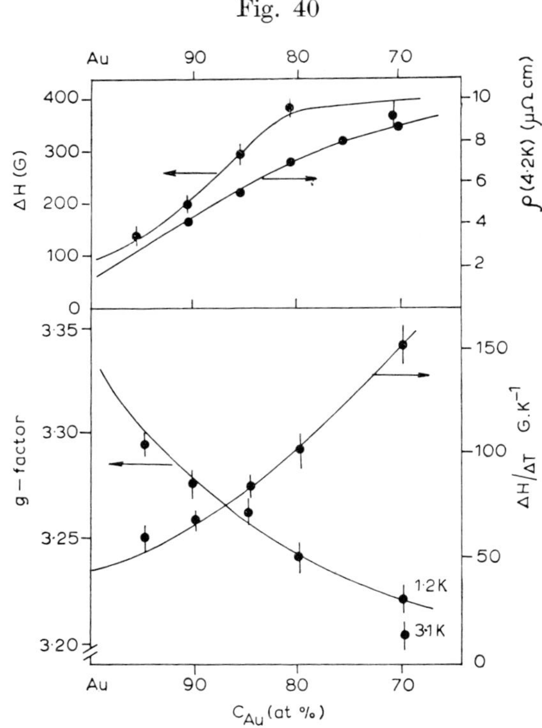

Summary of E.S.R. results for the Yb resonance in a Au_xAg_{1-x} host. The Yb concentration was 0·2%. Also shown is the residual resistivity, taken as the resistivity at 4·2 K (after Nagel *et al.*, A 100).

We conclude by mentioning the work of Nagel *et al.* (A 100) on the (Au_x Ag_{1-x}) Yb system. Since Yb is diamagnetic in Ag and paramagnetic in Au, the addition of Ag to Au allows the 3d virtual bound state to be drawn through the Fermi level. Whilst there is a negative g-shift for Yb in Au we would expect an even larger shift on adding Ag and indeed the authors observed such a shift; from a value of 3·34 for Yb in Au to 3·22 for Yb in $Au_{70}Ag_{30}$, with a corresponding increase in the slope of the linewidth from 40 G.K⁻¹ to 150 G.K⁻¹ (fig. 40). Both observations are in accord with a doubling of the exchange interaction. For Yb in Au, but increasingly as the Ag concentration was increased towards 30%, Nagel *et al.* observed a deepening low temperature minimum in the E.S.R. linewidth (fig. 41). The same alloys showed a clear Kondo minimum in their electrical resistance. The inference, that the Kondo effect is being observed directly in the E.S.R. linewidth is clear. At low concentrations of Yb ($\sim 0·2$ at. %), the broadening should not be due to magnetic ordering (although presumably the presence of chemical clustering or other inhomogeneity could modify this conclusion) but the absence of any clear temperature dependence of the g-value is rather worrying.

Fig. 41

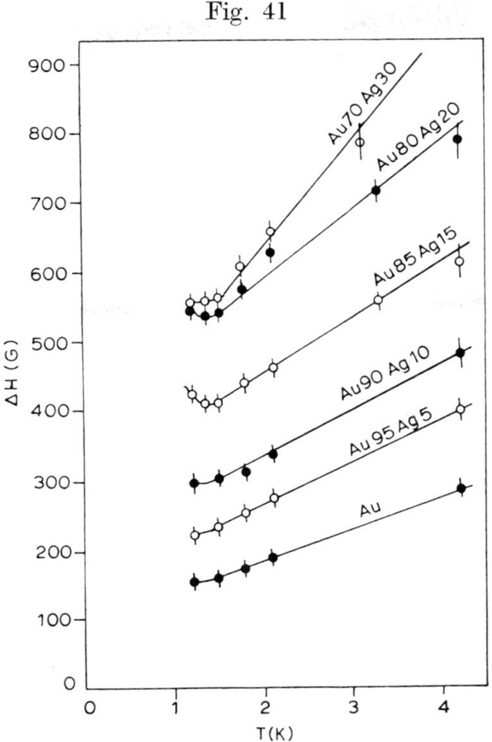

Temperature dependence of the linewidth of the Yb resonance in the system Yb : Au_xAg_{1-x} with a Yb concentration of 0·2% (A 100).

§ 5. Conclusions. The future direction of E.S.R. in metals

It would be extremely rash to try to predict new growth areas in E.S.R.; attempts at prediction of this sort always appear misguided within a few months; but what perhaps can be done is to examine the areas outlined in § 2 of this review and to try to estimate the extent of the problems still remaining.

In both the areas of the determination of exchange integrals and the study of the E.S.R. bottleneck, it is the author's opinion that the major part of the work has already been done. It would seem unlikely to be productive to continue simply to measure exchange integrals or demonstrate the existence of an E.S.R. bottleneck in increasingly exotic systems. This is not to say, however, that specific problems relating to the band structure of a particular alloy or compound or the resolution of a puzzling value for a g-shift or thermal broadening may not be useful—what it does imply is that the systematic collection of results for their own sake in these areas is unlikely to produce significant new developments. There are, however, a number of points which clearly do require to be resolved:

1. A clearer understanding of the reasons for the apparent discrepancy, between the measured g-values (which seem to be reliable and reproducible) for systems such as Pd–Gd and the results of high field magnetization measurements is needed.

2. A resolution of the long-standing debate about the origins of the negative contributions to the effective exchange. This is more likely to come from neutron-diffraction or N.M.R. studies rather than from E.S.R. measurements.

3. An answer to the question of why dynamic effects in the E.S.R. bottleneck are not consistently observed, in cases where they should be present.

4. Greater insight into the origins of the residual linewidth and in particular the role of the direct relaxation rate T_{dL}. Connected with this problem is the associated one of why an E.S.R. line has not been observed in a number of systems where one might expect it to be present, e.g. Rh–Mn, Pt–Mn, powder samples of Zn–Mn and the extreme breadth of the line in such cases as Ir–Gd.

5. The reasons for the negative residual linewidth which has been observed in a wide variety of compounds (e.g. $GdAl_2$, Pd_2MnSn) and dilute alloys (e.g. some Pd–Mn alloys).

6. Further work on double bottleneck systems—in particular the use of experiments such as those of Hirst $et\ al.$ (A 84) on Cu–Mn with Fe and Cr impurities, giving us information about the relaxation of these elements which, in general, we cannot observe directly.

In the general area concerned with the study of crystal-field effects in metals, E.S.R. has already made a notable contribution. Only recently have the signs of the fourth-order and sixth-order terms in the crystal-field Hamiltonian been studied for intermetallic compound systems and given that the problems involved in the preparation of good single crystals can be overcome, there remains much work to be done in this area. It would seem to the author that four areas may be of particular value in future research.

1. The recent observations of Davidov (A 57) and Baberschke $et\ al.$ (A 6) of a change in slope of the E.S.R. linewidth in Au : Er, where a low residual line-width enabled the resonance line to be followed to higher temperatures than is normally possible, suggests the possibility of direct estimates of the splitting of the crystal-field ground state from the first excited state to be made in other systems.

2. Likewise the recent observation of a resonance from the excited state in PdEr by Zingg and his co-workers (A 160) leads to the possibility, again, that similar effects may be observed in other systems.

3. The interesting conclusions of Moret $et\ al.$ (A 98) on the existence of the $M_s(\frac{1}{2} \leftrightarrow -\frac{1}{2})$ fine-structure transition in PdGd single crystals—where the authors have been able to interpret their results in terms of long-range spin–spin correlations—should encourage further measurements in related systems.

4. The recent observation of a resonance from Ce^{3+} in LaSb (A 49) and the speculation that a Tm resonance has been observed in YAl_2 (A 64) should provoke a search for resonance from these and other non-s-state rare earth ions in various hosts, particularly at the lowest temperatures.

The two areas in which the greatest potential contribution could be made by E.S.R. in the future are, in the author's opinion, the study of more concentrated alloys and of superconductivity. In the first of these areas a number of significant observations have already been made by E.S.R., including the observation of magnetic short-range ordering effects, field-cooling effects in spin-glasses, the observation of resonance from spiral phases, the use E.S.R. (together with more familiar techniques) to study phase transitions in metallic systems and to elucidate magnetic phase diagrams and in particular the study of magnetic order in both crystalline and amorphous materials (A 22) where high concentrations of impurity have been studied in sputtered thin films. The study of amorphous materials by E.S.R. may provide a promising new area of research.

The study of type II superconductors is fraught with problems, not least that of finding material where the critical field is sufficiently high to enable resonance to be observed in the superconducting phase. At present our theoretical understanding of resonance in superconductors is poor but as Orbach *et al.* (1973) have stressed that, if the theory can be made tractable, this area of research will have considerable potential. Orbach *et al.* compare the possibilities in the use of E.S.R. in superconductors with the present use of 'spin labels' in biology. Potentially the order parameter at a particular point in a material might be accessible directly through the shift and width of the E.S.R. line and the range of applicability might be extended by using the proximity effect to make super-conducting thin films of normal dilute alloys.

From the purely experimental point of view, many spectrometers currently in use are capable of detecting resonance from dilutions of magnetic impurities as low as 10–20 p.p.m. where the limiting factor is clearly the purity of the starting materials and 'cleanliness' of alloy preparation. One area in which experimental improvements may open new possibilities for E.S.R. in metals is in the construction of spectrometers capable of operation at temperatures below 0.5 K. The use of He^3–He^4 dilution refrigerators may enable temperatures as low as 30 mK to be achieved. At these temperatures there may be a real possibility of observing resonance from 3d impurities apart from Mn, and from Ce in a variety of hosts. Such an extension of the range of temperature may even provide much greater possibilities for the systematic study of the Kondo effect and of superconductivity.

 R. H. Taylor

REFERENCES

ABRAGAM, A., and BLEANEY, B., 1970, *Electron Paramagnetic Resonance of Transition Metals* (Oxford: Clarendon Press).
ABRIKOSOV, A. A., and GOR'KOV, L. P., 1962, *Soviet Phys. JETP*, **15**, 752.
ANDERSON, P. W., 1959, *Phys. Rev. Lett.*, **3**, 325; 1961, *Phys. Rev.*, **124**, 41.
ANDERSON, P. W., and WEISS, P. R., 1953, *Rev. mod. Phys.*, **25**, 269.
ASIK, J. R., BALL, M. A., and SLICHTER, C. P., 1969, *Phys. Rev.*, **181**, 645, 662.
BAGGULEY, D. M. S., CROSSLEY, W. A., and LIESEGANG, J., 1967, *Proc. phys. Soc.*, **90**, 1047.
BAGGULEY, D. M. S., and HEATH, M., 1967, *Proc. phys. Soc.*, **90**, 1029.
BARNES, S. E., 1974, *Phys. Rev.*, **9**, 4789.
BARNES, R. C., CORNELL, D. A., and TORGESON, D. R., 1966, *Phys. Rev. Lett.*, **16**, 233, shown to be spurious by GOSSARD, A. C., and WERNICK, J. H., 1966, *Phys. Rev. Lett.*, **16**, 955.
BARNES, S. E., DUPRAZ, J., and ORBACH, R., 1971, *J. appl. Phys.*, **42**, 1659.
BLEANEY, B., 1959, *Proc. phys. Soc.*, **73**, 939.
BUDWORTH, D. W., MOORE, F. E., and PRESTON, J., 1960, *Proc. R. Soc. A*, **257**, 25.
BUSCHOW, K. H. J., FAST, J. F., VAN DIEPEN, A. M., and DE WIJN, H. W., 1967, *Phys. Stat. Sol.*, **24**, 715.
CAMPBELL, I. A., 1972, *J. Phys. F*, **2**, 147.
CRANGLE, J., 1964, *Phys. Rev. Lett.*, **13**, 569.
DEBRAY, D. K., WALLACE, W. E., and RYBA, E., 1970, *J. less-common Metals*, **22**, 19.
DE GENNES, P. G., 1958, *C. r. hebd. Séanc. Acad. Sci., Paris*, **247**, 1836; 1962, *J. Phys., Paris*, **23**, 510.
DE WIJN, H. W., BUSCHOW, K. H. J., and VAN DIEPEN, A. M., 1968, *Phys. Stat. Sol.*, **30**, 759.
DIMMOCK, J. O., FREEMAN, A. J., and WATSON, R. E., 1966, *Proc. Int. Conf. on Optical Properties of Metals*, Amsterdam.
DIXON, J. M., and DUPREE, R., 1973, *J. Phys. F*, **3**, 118.
ELLIOTT, R. J., and STEVENS, K. W. H., 1952, *Proc. R. Soc. A*, **215**, 437; 1953, *Ibid.*, **218**, 533; **219**, 387.
FISCHER, K., 1971, *Phys. Stat. Sol. B*, **46**, 11.
FULDE, P., and MAKI, K., 1966, *Phys. Rev.*, **141**, 245.
GIOVANNINI, B., 1967, *Phys. Lett. A*, **26**, 80.
GOSSARD, A. C., JACCARINO, V., and WENICK, J. H., 1962, *Phys. Rev.*, **128**, 1038.
GÖTZE, W., 1973, *Proc. Conf. on EPR of Magnetic Ions in Metals*, Haute-Nendaz, p. 189.
GRÜNER, G., 1974, *Adv. Phys.*, **23**, 941.
GUERTIN, R. P., FONER. S., McNIFF, E. J., JR., and PRADDAUDE, H. C., 1971, *J. appl. Phys.*, **42**, 1550.
GUERTIN, R. P., PRADDAUDE, H. C., FONER, S., McNIFF, E. J., JR., and BARSOUMIAN, B., 1973, *Phys. Rev.*, *B*, **7**, 274.
HASEGAWA, H., 1959, *Prog. theor. Phys.*, **27**, 483.
HIRST, L. L., 1972, *Adv. Phys.*, **21**, 759.
HUTCHINGS, M., 1964, *Solid State Physics: Advances in Research and Applications*, Vol. 16, edited by F. Seitz and D. Turnbull (New York: Academic Press).
JHA, D., and JERICHO, M. H., 1971, *Phys. Rev. B*, **3**, 147.
KASUYA, J., 1956, *Prog. theor. Phys.*, **16**, 45.
KEFFER, P., and KITTEL, C., 1952, *Phys. Rev.*, **85**, 329.
KITTEL, C., 1948, *Phys. Rev.*, **73**, 155; 1968, *Solid State Physics: Advances in Research and Applications*, Vol. 22, edited by F. Seitz and D. Turnbull (New York: Academic Press), p. 1.
KONDO, J., 1969, *Solid State Physics* (New York: Academic Press), p. 23.
KORRINGA, J., 1950, *Physica, N.Y.*, **16**, 601.
KOUVEL, J. S., 1961, *J. Phys. Chem. Solids*, **21**, 57.
LEA, M. R., LEASK, M. J. M., and WOLF, W. P., 1962, *J. Phys. Chem. Solids*, **23**, 1381.
LIU, S. H., 1961, *Phys. Rev.*, **121**, 451.

Low, G. G., 1964, *Proc. Int. Conf. on Magnetism*, Nottingham (London: The Institute of Physics).
Maki, K., 1969, *Superconductivity*, edited by R. D. Parks (New York: Dekker), 2, 1035.
Monod, P., Hurdequint, H., Janossy, A., Obert, J., and Chaumont, J., 1975, *Phys. Rev. Lett.*, 18, 280.
Monod, P., and Schultz, S., 1968, *Phys. Rev.*, 173, 645.
Narath, A., and Weaver, H. R., 1968, *Phys. Rev.*, 175, 373.
Orbach, R., Peter, M., and Shaltiel, D., 1973, *Proc. Conf. on EPR of Magnetic Ions in Metals*, Haute-Nendaz, p. 9.
Orbach, R., and Spencer, H. J., 1968, *Physics Lett.* A, 26, 457.
Overhauser, A., 1953, *Phys. Rev.*, 89, 689.
Peter, M., Dupraz, J., and Cottet, H., 1967, *Helv. phys. Acta*, 40, 301.
Phillips, W. C., 1965, *Phys. Rev.*, 138, A1649.
Pifer, J. H., and Longo, R. T., 1971, *Phys. Rev.* B, 4, 3797.
Plefka, T., 1972, *Phys. Stat. Sol.* B, 51, K113.
Purwins, H. G., Walker, E., Donzé, P., Treyvaud, A., Furrer, A., Bührer, W., and Heer, H., 1973, *Solid St. Commun.*, 12, 117.
Rivier, N., 1974, *Wiss. Z. tech. Univ. Dresden* (to be published).
Rizzuto, C., 1974, *Rep. Prog. Phys.*, 37, 147.
Salamon, M. B., 1971, *Tech. Rep. 155P, 13B*, 1.
Schaller, H. J., Craig, R. S., and Wallace, W. E., 1972, *J. solid St. Chem.*, 5, 338
Schrieffer, J. R., and Wolff, P. A., 1966, *Phys. Rev.*, 194, 491.
Schultz, S., Shanabarger, M. R., and Platzman, P. M., 1967, *Phys. Rev. Lett.*, 19, 749.
Shull, C. G., and Yamada, Y., 1962, *J. phys. Soc. Japan*, Suppl. B111, 17, 1.
Spencer, H. J., and Doniach, S., 1967, *Phys. Rev. Lett.*, 18, 994.
Stewart, A. M., 1973, *J. Phys.* F, 3, 1024.
Sugawara, T., 1965, *J. phys. Soc. Japan*, 20, 2252.
Switendick, A. C., 1973, *Proc. 10th Rare Earth Research Conf.*, p. 235.
Taylor, K. N. R., 1971, *Adv. Phys.*, 20, 551.
Van Diepen, A. M., De Wijn, H. W., and Buschow, K. H. J., 1968, *Phys. Stat. Sol.*, 29, 187.
Walker, M. B., 1968, *Phys. Rev.*, 176, 432; 1973, *Proc. Conf. on EPR of Magnetic Ions in Metals*, Haute-Nendaz, p. 253.
Watson, R. E., Koide, S., Peter, M., and Freeman, A. J., 1965, *Phys. Rev.* A, 167, 139.
Williams, G., 1972, *J. Phys.* F, 2, L6.
Williams, G., and Hirst, L. L., 1969, *Phys. Rev.*, 185, 407.
Yafet, Y., 1968, *J. appl. Phys.*, 39, 853; 1973, *Proc. Conf. on EPR of Magnetic Ions in Metals*, Haute-Nendaz, p. 283.
Yosida, K., 1957, *Phys. Rev.*, 106, 893.
Zimmermann, P. H., Davidov, D., Orbach, R., Tao, L. J., and Zitkova, Z., 1972, *Phys. Rev.* B, 6, 2783.

APPENDIX

This appendix classifies E.S.R. experiments on magnetic impurities in metallic hosts by impurity and host element. The subjects to which an experiment contributes are shown by the numbers in the last column. These numbers are the subsection numbers of section 2 (pages 2–23) as follows:

1 g-shifts and temperature-dependent line widths; information about J_{eff}.
2 Information about the RKKY interaction.
3 The e.s.r. bottleneck.
4 Indentification of valence states.
5 Effect of the crystalline electric field.
6 Fine structure.
7 Hyperfine structure.
8 Magnetic ordering and related effects.
9 Superconductivity and e.s.r.
10 The Kondo effect.

The entry -ve in this column indicates that a resonance was sought but not found.

The Addendum to the Appendix (page 107) incorporates references up to 1 January 1976.

The papers quoted in this Appendix are listed as part 2 of the Appendix, those for the Addendum being numbered separately from A 161 onwards. References marked thus (*) and all those in the Addendum were received by the author too late for incorporation in the body of the text; those marked thus (†) describe work on single-crystal samples. Several papers quoted as part of the conference at Haute-Nendaz in 1973 have since been published in *Archives des Sciences (Genève)*, 1974, **27.**

Part 1. Classification by impurity and host

Impurity	Host	Other impurities	Year	Reference	Author	Subjects
Ce	LaSb	—	1974	(A 49)	Davidov *et al.*	5
*†	YSb	—	1974	(A 59)	Davidov *et al.*	5
Dy	Ag	—	1966	(A 79)	Griffiths, Coles	-ve
	Ag	—	1968	(A 83)	Hirst *et al.*	-ve
	Ag	—	1971	(A 43)	Davidov *et al.*	5
†	Ag	—	1973	(A 57)	Davidov *et al.*	1, 5, 7
	Al	—	1973	(A 122)	Rettori *et al.*	1, 5, 7
†	Au	—	1972	(A 41)	Davidov *et al.*	1, 5
	Ir	—	1972	(A 50)	Davidov *et al.*	1, 7
†	Pd	—	1972	(A 63)	Devine *et al.*	1, 5, 7
†	Pd	—	1973	(A 62)	Devine	5
	Rh	—	1971	(A 53)	Davidov *et al.*	1, 7
	Th	—	1972	(A 40)	Davidov *et al.*	1, 7
	LaSb	—	1974	(A 49)	Davidov *et al.*	5
†	YAl	—	1973	(A 64)	Devine *et al.*	1, 5
*†	YSb	—	1974	(A 59)	Davidov *et al.*	5
	RCO_2	—	1972	(A 18)	Burzo, Domsa	—
	RCO_2	—	1972	(A 17)	Burzo	—
Er	Ag	—	1966	(A 79)	Griffiths, Coles	5
	Ag	—	1968	(A 83)	Hirst *et al.*	1, 5
	Ag	—	1970	(A 26)	Chui *et al.*	4, 7
	Ag	—	1973	(A 6)	Baberschke	5
	Ag	—	1973	(A 57)	Davidov *et al.*	1
*	Ag	—	1975	(A 5)	Arbeli *et al.*	Thin films
	Al	—	1973	(A 122)	Rettori *et al.*	1, 5, 7
	Au	—	1963	(A 83)	Hirst *et al.*	1, 5
	Au	—	1971	(A 146)	Tao *et al.*	1, 7
	Au	—	1972	(A 25)	Chock *et al.*	7
	Au	—	1973	(A 57)	Davidov *et al.*	1, 5, 7

Part 1 (*Continued*)

		Au	—	1973	(A 6)	Baberschke *et al.*	
		Cu	—	1971	(A 53)	Davidov *et al.*	1, 5
	*	Cu	—	1974	(A 3)	Al'tshuler *et al.*	1, 4
		Ir	—	1972	(A 50)	Davidov *et al.*	1, 7
		La	—	1973	(A 1)	Alekseevskii *et al.*	9
	†	Mg	—	1967	(A 10)	Burr and Orbach	1, 3
		Pd	H	1972	(A 42)	Davidov *et al.*	1, 5, 7
		Pd	—	1972	(A 67)	Devine *et al.*	1, 5, 7
	†	Pd	—	1974	(A 160)	Zingg *et al.*	5
		Pt	—	1971	(A 53)	Davidov *et al.*	1, 5
		Pt	—	1972	(A 42)	Davidov *et al.*	1, 5, 7
		Rh	—	1971	(A 53)	Davidov *et al.*	1, 5
		Rh	—	1972	(A 42)	Davidov *et al.*	1, 7
	†	Sc	—	1974	(A 86)	Keller *et al.*	5
Er		Th	—	1971	(A 53)	Davidov *et al.*	1, 5
		Th	—	1972	(A 40)	Davidov *et al.*	1, 7
	*	Th	—	1974	(A 125)	Rettori, Davidov	1, 7
	†	YAl_2	—	1973	(A 64)	Devine *et al.*	1, 5
	†	LaBi	—	1974	(A 49)	Davidov *et al.*	5
	†	LuBi	—	1974	(A 49)	Davidov *et al.*	5, 7
	†	LaSb	—	1974	(A 49)	Davidov *et al.*	5
	†	LuSb	—	1974	(A 49)	Davidov *et al.*	5
		LaB_6	—	1974	(A 49)	Davidov *et al.*	5
	*†	YSb	—	1974	(A 59)	Davidov *et al.*	5
		RCO_2	—	1972	(A 18)	Burzo, Domsa	-ve
		RCO_2	—	1972	(A 17)	Burzo	-ve
Eu		Ca	—	1972	(A 133)	Schmidt	1, 4
		Eu	—	1960	(A 113)	Peter, Matthias	3, 8
		Eu	—	1971	(A 69)	Ehara, Arrott	1, 8
	*	La	—	1975	(A 92)	Koopmann *et al.*	1, 4
		Mg	—	1967	(A 10)	Burr, Orbach	1
		Y	—	1971	(A 69)	Ehara, Arrott	1
		Yb	—	1968	(A 129)	Schäfer *et al.*	1, 3, 8
		Yb	—	1969	(A 131)	Schäfer *et al.*	1, 3, 8

Part 1 (*Continued*)

	Yb	—	1971	(A 69)	Ehara, Arrott	1
	Yb	Ca	1972	(A 134)	Schmidt *et al.*	1, 4
	Yb	—	1972	(A 133)	Schmidt	1, 4
	LaAl$_2$	—	1972	(A 90)	Koopmann *et al.*	1, 4
	BaAl$_4$	—	1973	(A 72)	Elschner *et al.*	1, 6
	RAl$_2$	—	1960	(A 85)	Jaccarinno *et al.*	1, 2
	RAl$_2$	—	1961	(A 112)	Peter	3
	RAl$_2$	—	1975	(A 151)	Taylor, Coles	1, 8
	RAl$_4$	—	1967	(A 159)	Wernick *et al.*	1, 3, 8
	RAl$_4$	—	1973	(A 148)	Taylor, Coles	1, 8
	RAl$_4$	—	1975	(A 151)	Taylor, Coles	1, 8
	RIr$_2$	—	1960	(A 113)	Peter, Matthias	-ve
	RMg$_2$	—	1961	(A 112)	Peter	3
	RPd	—	1975	(A 151)	Taylor, Coles	1, 8
	RPd$_2$	—	1975	(A 151)	Taylor, Coles	1, 8
Fe	Al	—	1962	(A 30)	Collings, Hedgcock	-ve
	Cr	—	1968	(A 128)	Salamon, Feigl	
	Mg	—	1962	(A 30)	Collings, Hedgcock	-ve
	Ti	—	1970	(A 11)	Burr *et al.*	1, 8
	V	—	1970	(A 11)	Burr *et al.*	1, 8
	CeRu$_2$	—	1973	(A 91)	Koopmann *et al.*	1
Gd	Ag	—	1962	(A 115)	Peter *et al.*	1
	Ag	—	1966	(A 79)	Griffiths, Coles	1
	Ag	—	1968	(A 83)	Hirst *et al.*	1
	Ag	Al	1970	(A 104)	Nakamura *et al.*	1, 4
†	Ag	Au	1973	(A 57)	Davidov *et al.*	1
	Ag	—	1974	(A 22)	Charles *et al.*	1, 8
	Al	Ag, Au, Pt	1973	(A 122)	Rettori *et al.*	1, 4
†	Au	—	1971	(A 24)	Chock *et al.*	7
	Cd	—	1962	(A 115)	Peter *et al.*	1

Part 1 (*Continued*)

	Er	—	1965	(A 117)	Pop, Chechernikov	1, 8
	Gd	—	1953	(A 87)	Kip	1, 8
†	Gd	—	1964	(A 126)	Rodbell, Moore	1, 2, 8
	Gd	—	1966	(A 82)	Harris *et al.*	1
	Gd	—	1970	(A 23)	Chiba, Nakamura	1
	Gd	—	1972	(A 16)	Burzo, Domsa	1, 2, 8
	Ir	—	1972	(A 50)	Davidov *et al.*	—
	La	—	1966	(A 82)	Harris *et al.*	1, 8
	La	—	1971	(A 35)	Cottet	1, 4
*	La	—	1975	(A 92)	Koopman *et el.*	1, 4
	Lu	—	1965	(A 81)	Harris *et al.*	1, 8
	Lu	—	1966	(A 82)	Harris *et al.*	1, 8
	Mg	—	1967	(A 10)	Burr, Orbach	1
	Mg	—	1971	(A 68)	Ehara	1
†	Mg	—	1971	(A 147)	Tao *et al.*	6
	Pd	Ag, Cd, Rh	1962	(A 115)	Peter *et al.*	1, 2, 8
	Pd	La, Ce, Pr, Nd / Lu, Tb, Dy, Ho / Er, Tm, Yb	1962	(A 114)	Peter *et al.*	2
	Pd	La, Ce, Pr, Nd, Lu / Tb, Dy, Ho, Er, Tm / Yb, Rh, Ag, Fe, Co, / Ni, U	1964	(A 140)	Shaltiel *et al.*	1, 2, 8
	Pd	—	1965	(A 81)	Harris *et al.*	1, 8
	Pd	Rh, Ni, Ag, Cr, Pt	1970	(A 32)	Cottet, Peter	1
	Pd	—	1970	(A 28)	Coles *et al.*	1
	Pd	Rh, Ni, Ag, Cr, Pt	1971	(A 35)	Cottet	1
	Pd	—	1971	(A 121)	Praddaude, Gärtner	1
†	Pd	—	1972	(A 65)	Devine *et al.*	1, 5
	Pd	—	1972	(A 42)	Davidov *et al.*	1
	Pd	Ni, Pt	1974	(A 150)	Taylor, Coles	1, 4, 8
†	Pd	—	1974	(A 98)	Moret *et al.*	1, 2, 6
Gd	PdH	—	1963	(A 136)	Shaltiel	1
	Pt	—	1962	(A 116)	Peter *et al.*	-ve
	Pt	—	1972	(A 42)	Davidov *et al.*	1

Part 1 (*Continued*)

	Rh	—	1962	(A 116)	Peter *et al.*	1
	Rh	—	1971	(A 53)	Davidov *et al.*	1
	Rh	—	1972	(A 42)	Davidov *et al.*	1
	Sc	—	1966	(A 82)	Harris *et al.*	1, 8
†	Sc	—	1971	(A 127)	Salamon	1, 6
	Sc	—	1971	(A 35)	Cottet	1, 4
	Tb	—	1965	(A 117)	Pop, Chechernikov	1, 8
	Th	—	1972	(A 40)	Davidov *et al.*	1
	Y	—	1963	(A 118)	Popplewell, Tebble	1, 2, 8
	Y	—	1965	(A 81)	Harris *et al.*	1, 8
	Y	—	1965	(A 117)	Pop, Chechernikov	1, 8
	Y	—	1966	(A 82)	Harris *et al.*	1, 8
	Y	—	1971	(A 35)	Cottet	1, 4
†	Y	—	1971	(A 71)	Elschner, Weimann	1, 4, 6
†	Y	—	1973	(A 158)	Weimann, Elschner	6, 8
†	Y	—	1974	(A 9)	Bagguley *et al.*	8
	Rh–Ni	—	1968	(A 34)	Cottet *et al.*	1
	LaAg	—	1973	(A 48)	Davidov *et al.*	1
	YAg	—	1972	(A 157)	Weimann *et al.*	1
	YAg	Th	1973	(A 156)	Weimann *et al.*	1, 4
	YAg	—	1973	(A 48)	Davidov *et al.*	1
	RAg	—	1972	(A 19)	Burzo *et al.*	1, 2, 8
	RAg₃	—	1973	(A 148)	Taylor, Coles	1, 8
	RAg₃	—	1975	(A 151)	Taylor, Coles	1, 8
	LaAl₂	—	1964	(A 139)	Shaltiel *et al.*	1
	LaAl₂	—	1970	(A 28)	Coles *et al.*	1
	LaAl₂	—	1972	(A 90)	Koopmann *et al.*	1, 4
	LaAl₂	Ce, Th, U	1971	(A 47)	Davidov *et al.*	4
	LaAl₂	Th	1973	(A 56)	Davidov *et al.*	1, 4, 9
	LaAl₂	—	1973	(A 149)	Taylor	1, 4, 8
	YAl₂	—	1970	(A 130)	Schäfer *et al.*	1, 4
	YAl₂	Th	1972	(A 132)	Schäfer *et al.*	1, 4

Part 1 (*Continued*)

	YAl_2	—	1974	(A 124)	Rettori *et al.*	1, 4
	$LuAl_2$	Th	1974	(A 124)	Rettori *et al.*	1, 4
	RAl_2	—	1960	(A 85)	Jaccarino *et al.*	1, 2
	RAl_2	Pr, Nd, Sm, Tb, Dy, Er	1961	(A 112)	Peter	1, 2
	RAl_2	—	1968	(A 44)	Shaltiel, Davidov	1
	RAl_2	—	1972	(A 80)	Hacker *et al.*	1
	RAl_2	—	1975	(A 151)	Taylor, Coles	1, 8
	RAu	—	1973	(A 14)	Burzo, Baican	1, 2
	LaB_6	—	1973	(A 142)	Sperlich *et al.*	1, 2, 8
	SrB_6	—	1974	(A 89)	Kojima *et al.*	1
Gd	LaB_6	—	1974	(A 89)	Kojima *et al.*	1
	SmB_6	—	1974	(A 89)	Kojima *et al.*	1
	RB_6	—	1962	(A 27)	Coles *et al.*	1
	RB_6	—	1971	(A 95)	Miller, Hacker	1
	RB_6	—	1971	(A 74)	Fisk *et al.*	1, 8
	RB_6	—	1972	(A 141)	Sperlich, Janneck	2, 8
	RB_6	—	1973	(A 143)	Sperlich	1, 2, 8
	RB_6	—	1973	(A 148)	Taylor, Coles	1, 8
	RB_6	—	1971	(A 74)	Fisk *et al.*	1, 8
	RB_6	—	1973	(A 148)	Taylor, Coles	1, 8
*†	$LaBi$	—	1974	(A 152)	Urban *et al.*	1, 6
	RCO_2	—	1972	(A 17)	Burzo	1, 8
	RCO_2	—	1972	(A 18)	Burzo, Domsa	1, 8
	YCu	—	1973	(A 48)	Davidov *et al.*	1
	RCu	—	1972	(A 19)	Burzo *et al.*	1, 2, 8
*	YCu	—	1974	(A 15)	Burzo, Baican	1, 2, 4
	RCu_2	—	1973	(A 148)	Taylor, Coles	1, 8
	RCu_2	—	1975	(A 151)	Taylor, Coles	1, 8
	RCu_4	—	1973	(A 148)	Taylor, Coles	1, 8
	RCu_4	—	1975	(A 151)	Taylor, Coles	1, 8
	YCu_5	—	1964	(A1 40)	Shaltiel *et al.*	1, 8
	RCu_5	—	1964	(A 140)	Shaltiel *et al.*	1, 8
	RCu_6	—	1968	(A 108)	Okuda, Date	8
	RCu_6	—	1975	(A 151)	Taylor, Coles	1, 8

Part 1 (*Continued*)

	La$_3$In	—	1972	(A 4)	Al'tshuler *et al.*	1, 9
	LaIr$_2$	—	1964	(A 139)	Shaltiel *et al.*	1
	RIr$_2$	—	1960	(A 113)	Peter, Matthais	8
	RIr$_2$	—	1968	(A 44)	Davidov, Shaltiel	1
	ThIr$_5$	—	1964	(A 140)	Shaltiel *et al.*	1
	RMn$_2$	—	1968	(A 44)	Davidov, Shaltiel	1
	RNi	—	1972	(A 153)	Ursu, Burzo	1, 2
	RNi$_2$	—	1972	(A 13)	Burzo, Laforest	1, 2
	RNi$_2$	—	1972	(A 153)	Ursu, Burzo	1, 2
	LaNi$_5$	—	1964	(A 140)	Shaltiel *et al.*	1
	LaNi$_5$	—	1968	(A 45)	Davidov, Shaltiel	1, 4
	LaNi$_5$	—	1975	(A 93)	Male, Taylor	1, 4, 5
	ThNi$_5$	—	1964	(A 140)	Shaltiel *et al.*	1
	YNi$_5$	Pr, Tb, Er	1964	(A 140)	Shaltiel *et al.*	1
	UNi$_5$	—	1964	(A 140)	Shaltiel *et al.*	1, 8
	RNi$_5$	—	1964	(A 140)	Shaltiel *et al.*	1, 8
	RNi$_5$	—	1971	(A 12)	Burzo, Ursu	1, 2
	RNi$_5$	—	1972	(A 153)	Ursu, Burzo	1, 2
*	LaOs$_2$	—	1975	(A 135)	Schrittenlacher *et al.*	9
	LaPt$_2$	—	1964	(A 139)	Shaltiel *et al.*	1
	RPt$_2$	—	1968	(A 145)	Vijaraghavan *et al.*	1
Gd	RPt$_2$	—	1968	(A 155)	Vijaraghavan *et al.*	1
	RPt$_2$	—	1968	(A 44)	Davidov, Shaltiel	1
	RPt$_{2/3}$	—	1975	(A 151)	Taylor, Coles	1, 8
	LaPt$_5$	—	1964	(A 140)	Shaltiel *et al.*	1
	RPd$_3$	—	1975	(A 151)	Taylor, Coles	1, 8
	LaPd$_3$	—	1974	(A 49)	Davidov *et al.*	1
	UPd$_3$	—	1964	(A 140)	Shaltiel *et al.*	1, 8
	UThPd$_3$	—	1969	(A 37)	Davidov *et al.*	10
	LaRh$_2$	—	1964	(A 139)	Shaltiel *et al.*	1
	RRh$_2$	—	1968	(A 44)	Davidov, Shaltiel	1

R. H. Taylor

Part 1 (*Continued*)

	RRh$_2$	—	1975	(A 151)	Taylor, Coles	1, 8
	CeRu$_2$	—	1975	(A 8)	Baberschke *et al.*	1, 9
	CeRu$_2$	—	1964	(A 139)	Shaltiel *et al.*	1, 9
	CeRu$_2$	—	1971	(A 142)	Sun, Schnitzke	1, 8
	CeRu$_2$	—	1973	(A 73)	Engel *et al.*	9
	CeRu$_2$	—	1973	(A 46)	Davidov *et al.*	1, 9
	CeRu$_2$	—	1974	(A 7)	Baberscke *et al.*	1, 9
	CeRu$_2$	—	1974	(A 51)	Davidov *et al.*	1, 9
	LaRu$_2$	—	1964	(A 139)	Shaltiel *et al.*	1
	LaRu$_2$	Th	1965	(A 137)	Shaltiel *et al.*	1
	LaRu$_2$	Tb, Pr	1968	(A 33)	Cottet *et al.*	1, 4, 8
	LaRu$_2$	Pr	1969	(A 38)	Davidov *et al.*	1, 4
	LaRu$_2$	—	1973	(A 123)	Rettori *et al.*	1, 9
	LaRu$_2$	—	1973	(A 73)	Engel *et al.*	9
	LaRu$_2$	—	1973	(A 46)	Davidov *et al.*	1, 9
	LaRu$_2$	—	1974	(A 51)	Davidov *et al.*	9
	LaRu$_2$	—	1975	(A 8)	Baberschke *et al.*	9
	ScRu$_2$	—	1964	(A 139)	Shaltiel *et al.*	1
	ThRu$_2$	—	1964	(A 139)	Shaltiel *et al.*	1, 9
	ThRu$_2$	—	1973	(A 46)	Davidov *et al.*	1, 9
	ThRu$_2$	—	1974	(A 52)	Davidov *et al.*	1, 9
	ThRu$_2$	—	1974	(A 51)	Davidov *et al.*	1, 9
	YRu$_2$	—	1964	(A 139)	Shaltiel *et al.*	1
	ZrRu$_2$	—	1964	(A 139)	Shaltiel *et al.*	1
*†	LaSb	—	1974	(A 55)	Davidov *et al.*	6
*†	PrSb	—	1975	(A 58)	Davidov *et al.*	5, 6
*†	YSb	—	1974	(A 59)	Davidov *et al.*	6
	ZrZn$_2$	—	1971	(A 36)	Davidov *et al.*	1, 4
	RZn$_2$	—	1971	(A 60)	Debray, Ryba	1, 2
	RZn$_{12}$	—	1973	(A 148)	Taylor, Coles	1, 8
	RZn$_{12}$	—	1975	(A 151)	Taylor, Coles	1, 8
	R$_2$Zn$_{17}$	—	1973	(A 148)	Taylor, Coles	1, 8
	R$_2$Zn$_{17}$	—	1975	(A 151)	Taylor, Coles	1, 8
Ho	Ag	—	1966	(A 79)	Griffiths, Coles	-ve
	RCO$_2$	—	1972	(A 18)	Burzo, Domsa	-ve

Part 1 (*Continued*)

		RCO$_2$	—	1972	(A 17)	Burzo	-ve
	Mn	Ag	—	1957	(A 110)	Owen *et al.*	1, 8
		Ag	—	1964	(A 138)	Shaltiel, Wernick	1
		Ag	Zn, Ga, Cd, Pt, Au	1967	(A 77)	Gossard *et al.*	4
		Ag	—	1969	(A 109)	Okuda, Date	8
		Ag	—	1970	(A 70)	Elliston	1, 4, 8
		Ag	—	1971	(A 99)	Nagashima, Abe	4, 8
		Ag	Au, Sb	1975	(A 54)	Davidov *et al.*	1, 4, 10
		Al	—	1962	(A 30)	Collings, Hedgcock	-ve
		Au	—	1964	(A 138)	Shaltiel, Wernick	1
		Au	V, Fe, Ni	1970	(A 106)	Oda, Asayama	4
		Cu	—	1956	(A 111)	Owen *et al.*	1
†		Cu	—	1957	(A 110)	Owen *et al.*	1, 8
		Cu	—	1960	(A 144)	Street	8
†		Cu	—	1966	(A 75)	Geschwind *et al.*	7
		Cu	Ti, Fe, Ga, Ni, Al, Mg	1967	(A 76)	Gossard *et al.*	1, 4
		Cu	—	1967	(A 78)	Griffiths	8
		Cu	—	1967	(A 101)	Nakamura, Kinoshita	4, 8
		Cu	—	1967	(A 102)	Nakamura, Kinoshita	4
		Cu	Al, Ni	1967	(A 107)	Okuda, Date	4, 8
		Cu	Ni, Co, Fe	1968	(A 105)	Oda, Asayama	4
		Cu	Al, Si	1968	(A 94)	McElroy, Heeger	1, 4
		Cu	Al, Zn, Ti, Pd Ni, Fe, Co	1969	(A 109)	Okuda, Date	4, 8
		Cu	—	1969	(A 103)	Nakamura, Kinoshita	1, 4, 8
		Cu	—	1969	(A 96)	Miyako *et al.*	8
		Cu	Fe, Co, Ni	1970	(A 106)	Oda, Asayama	1, 4, 10
		Cu	—	1971	(A 99)	Nagashima, Abe	4, 8
		Cu	Cr	1973	(A 84)	Hirst *et al.*	4, 5
†		Mg	—	1957	(A 110)	Owen *et al.*	1
		Mg	—	1962	(A 30)	Collings, Hedgcock	1, 8

Part 1 (*Continued*)

	Mg	—	1971	(A 88)	Kleinhams, Wigen	1, 5, 8
	Pd	—	1964	(A 140)	Shaltiel *et al.*	1
	Pd	Pr, Ho, Tb	1964	(A 138)	Shaltiel, Wernick	1, 2
	Pd	—	1971	(A 35)	Cottet	1, 4
	Pd	—	1974	(A 2)	Alquié, *et al.*	1, 8
	Pd	Au, Pt	1975	(A 29)	Coles *et al.*	1, 4, 8
	PdH	—	1974	(A 2)	Alquié *et al.*	1, 8
	Zn	—	1962	(A 31)	Collings *et al.*	1, 8
	Zn	Al, Cu	1970	(A 97)	Miyako	1
†	Zn	—	1974	(A 61)	Devine, Moret	1, 4
	$Zn_{13}T$	Fe	1971	(A 20)	Caplin *et al.*	8
	$Zn_{13}T$	Fe	1974	(A 21)	Caplin *et al.*	2, 8
	$Zn_{13}T$	—	1975	(A 151)	Taylor, Coles	2, 8
	$Zn_{10}T$	—	1975	(A 151)	Taylor, Coles	8
	$MoGa_4$	—	1972	(A 66)	Devine *et al.*	4, 8, 10
Tm	Ag	—	1968	(A 83)	Hirst *et al.*	-ve
†	YAl_2	—	1973	(A 64)	Devine *et al.*	5
						possible
Tb	RCo_2	—	1972	(A 18)	Burzo, Domsa	—
Yb	Au	—	1968	(A 83)	Hirst *et al.*	1, 5
	Au	—	1971	(A 146)	Tao *et al.*	1, 7
	Au	—	1972	(A 25)	Chock *et al.*	7
	Au : Ag	—	1973	(A 100)	Nagel *et al.*	1, 10
	$LaPd_3$	—	1974	(A 49)	Davidov *et al.*	1, 5, 7
	LaSb	—	1974	(A 49)	Davidov *et al.*	5
*†	YSb	—	1974	(A 59)	Davidov *et al.*	5

Addendum to Appendix (to 1 January 1976)

Impurity	Host	Other Impurities	Year	Reference	Author	Subjects
Dy	Pd	—	1975	(A 171)	Praddaude	5
	Pd	—	1975	(A 178)	Zingg et al.	1,5
	LaSb	—	1975	(A 165)	Davidov et al.	5
	ScAl$_2$	—	1975	(A 167)	Devine et al.	1,5
Er	Ag	—	1975	(A 161)	Arbilly et al.	epitaxial films
	Au	—	1975	(A 176)	Sjostrand, Seidel	1,5,7
	Pd	—	1975	(A 178)	Zingg et al.	1,5
	LaSb	—	1975	(A 174)	Rettori et al.	1,5
	LuAl$_2$	—	1975	(A 173)	Rettori et al.	1,3,5
	PrSb	—	1975	(A 174)	Rettori et al.	1,5
	ScAl$_2$	—	1975	(A 167)	Devine et al.	1,5
	TmSb	—	1975	(A 174)	Rettori et al.	1,5
	YSb	—	1975	(A 174)	Rettori et al.	1,5
Gd	Ag	—	1975	(A 170)	Oseroff et al.	1
	Pd	Sm	1975	(A 169)	Malik, Vijarajhavan	1,5
	Tb	—	1975	(A 162)	Bajguley, Partington	8
	LaSb	—	1975	(A 177)	Urban et al.	1,5,6
	LuAl$_2$	Ce	1975	(A 173)	Rettori et al.	1,3,5
	PrBi	—	1975	(A 172)	Rettori et al.	1,5,8

Addendum to Appendix *(Continued)*

Gd	PrSb	—	1975	(A 172)	Rettori *et al.*	1,5,8
	PrTe	—	1975	(A 172)	Rettori *et al.*	1,5,8
	ScAg	—	1975	(A 175)	Seipler, Elschner	1
	ScCu	—	1975	(A 175)	Seipler, Elschner	1
	ScPd	—	1975	(A 175)	Seipler, Elschner	1
	ScRh	—	1975	(A 175)	Seipler, Elschner	1
	ScRu	—	1975	(A 175)	Seipler, Elschner	1
	TmSb	—	1975	(A 164)	Davidov, Baberschke	1,5
	TmBi	—	1975	(A 172)	Rettori *et al.*	1,5,8
	TmSb	—	1975	(A 172)	Rettori *et al.*	1,5,8
	YAl_2	Ce	1975	(A 173)	Rettori *et al.*	1,3,5
	YRh	—	1975	(A 175)	Seipler, Elschner	1
	$GdFe_2$	—	1975	(A 163)	Bhajat, Paul	amorphous
Mn	Mg	—	1975	(A 168)	Kleinhams *et al.*	8
	Pd	—	1975	(A 166)	Devine *et al.*	1,8
	PdH	—	1975	(A 166)	Devine *et al.*	1,8
Tb	$TbFe_2$	—	1975	(A 163)	Bhajat, Paul	amorphous

Tb	RCo_2	—	1972	(A 18)	Burzo, Domsa	

Part 2. Classification by authors

(A 1) ALEKSEEVSKII, N. E., GARAFULLIN, I. A., KOCHALAEV, B. I., and KARA-KHASH'YAN, E. G., 1973, *Zh. éksp. teor. Fiz. Pis. Red.*, **18**, 323.

(A 2) ALQUIÉ, G., KREISLER, A., SADOC, G., and BURGER, J. P., 1974, *J. Phys. Lett.*, **35**, 69.

(A 3)* AL'TSHULER, T. S., ZARIPOV, M. M., KUKOVITSKII, E. F., KHAIMOVICH, E. P., and KHARAKHASH'YAN, E. G., 1974, *Zh. éskp teor. Fiz. Pis'ma (USSR)*, **20**, 413; 1975, *J. exp. theor. Phys. Lett.*, **20**, 349.

(A 4) AL'TSHULER, T. S., GAROFULLIN, I. A., and KHARAKHASH'YAN, E. G., 1972, *Soviet Phys. solid St.*, **14**, 213.

(A 5)* ARBELI, D., DEUTSCHER, G., GRUNBAUM, E., SUSS, J.T., and ORBACH, R., 1975 (to be published).

(A 6) BABERSCHKE, K., 1973, *Proc. Conf. on ESR of Ions in Metals* (Haute Nendaz), p. 111.

(A 7) BABERSCHKE, K., ENGEL, U., and HÜFNER, S., 1974, *Solid St. Commun.*, **15**, 1101.

(A 8) BABERSCHKE, K., ENGEL, U., KOOPMANN, G., and HÜFNER, S., *AIP Conf. Proc.*(18), Boston, U.S.A., p. 84.

(A 9) BAGGULEY, D. M. S., LIESEGANG, J., and ROBINSON, K., 1974, *J. Phys. F*, **4**, 594.

(A 10) BURR, C. R., and ORBACH, R., 1967, *Phys. Rev. Lett.*, **19**, 1133.

(A 11) BURR, C. R., ZINGG, W., and PETER, M., 1970, *Helv. phys. Acta*, **43**, 771.

(A 12) BURZO, E., and URSU, I., 1971, *Solid St. Commun.*, **9**, 2289.

(A 13) BURZO, E., and LAFOREST, J., 1972, *Int. J. Magn.*, **3**, 171.

(A 14) BURZO, E., and BAICAN, R., 1973, *Colloque Ampère*, **XVII**, 517.

(A 15)* BURZO, E., and BAICAN, R., 1974, *Colloque Ampère*, **XVIII**, p. 337

(A 16) BURZO, E., and DOMSA, F., 1972, *Rev. roum. Phys.*, **17**, 893.

(A 17) BURZO, E., 1972, *Int. J. Magn.*, **3**, 161.

(A 18) BURZO, E., and DOMSA, F., 1972, *Phys. Stat. Sol. B*, **51**, K89.

(A 19) BURZO, E., URSU, I., and PIERRE, J., 1972, *Phys. Stat. Sol. B*, **51**, 463.

(A 20) CAPLIN, D., DUNLOP, J. B., and TAYLOR, R. H., 1971, *J. Phys.*, 52–3, 32, C16.

(A 21) CAPLIN, D., and DUNLOP, J. B., 1974, *J. Phys. F*, **3**, 1621.

(A 22) CHARLES, S. W., POPPLEWELL, J., and BATES, P. A., 1974, *J. Phys. F*, **3**, 664.

(A 23) CHIBA, Y., and NAKAMURA, A., 1970, *J. phys. Soc. Japan*, **29**, 792.

(A 24) CHOCK, E. P., CHUI, R., DAVIDOV, D., ORBACH, R., SHALTIEL, D., and TAO, L. J., 1971, *Phys. Rev. Lett.*, **27**, 582.

(A 25) CHOCK, E. P., DAVIDOV, D., ORBACH, R., RETTORI, C., and TAO, L. J., 1972, *Phys. Rev. B*, **5**, 2735.

(A 26) CHUI, R., ORBACH, R., and GEHMAN, B. L., 1970, *Phys. Rev. B*, **2**, 2298.

(A 27) COLES, B. R., COLE, T., LAMBE, J., and LAURENCE, N., 1962, *Proc. Phys. Soc.*, **79**, 84.

(A 28) COLES, B. R., GRIFFITHS, D., LOWIN, R. J., and TAYLOR, R. H., 1970, *J. Phys. C*, **3**, L121.

(A 29) COLES, B. R., JAMIESON, H., TAYLOR, R. H., and TARI, A., 1975, *J. Phys. F*, **5**, 565.

(A 30) COLLINGS, E. W., and HEDGCOCK, F. T., 1962, *Phys. Rev.*, **126**, 1654.

(A 31) COLLINGS, E. W., HEDGCOCK, F. T., and MUIR, W. B., 1962, *Proc. 8th Conf. on Low Temp. Phys.*, London, p. 253.

(A 32) COTTET, H., and PETER, M., 1970, *Solid St. Commun.*, **8**, 1601.

(A 33) COTTET, H., DONZÉ, P., ORTELLI, J., WALKER, E., and PETER, M., 1968, *Helv. phys. Acta*, **41**, 755.

(A 34) COTTET, H., DONZÉ, P., DUPRAZ, J., GIOVANNINI, B., and PETER, M., 1968, *Z. angew. Phys.*, **24**, 249.

(A 35) COTTET, H., 1971, Doctoral Thesis, University of Geneva (unpublished).

(A 36) DAVIDOV, D., DOUBLON, G., and SHALTIEL, D., 1971, *Phys. Rev. B*, **3**, 3651.

(A 37)   DAVIDOV, D., LOTEM, H., SHALTIEL, D., WEGER, M., and WERNICK, J. H.,
          1969, *Phys. Rev.*, **184**, 481.
(A 38)   DAVIDOV, D., LOTEM, H., and SHALTIEL, D., 1969, *Physics Lett.* A, **28**, 672.
(A 39)   DAVIDOV, D., RETTORI, C., ORBACH, R., DIXON. A., and CHOCK, E. P., 1975,
          (to be published).
(A 40)   DAVIDOV, D., ORBACH, R., RETTORI, C., SHALTIEL, D., TAO, L. J., and
          RICKS, B., 1972, *Phys. Rev.* B, **5**, 1711.
(A 41)   DAVIDOV, D., ORBACH, R., RETTORI, C., TAO, L. J., and CHOCK, E. P., 1972,
          Phys. Rev. Lett., **28**, 490.
(A 42)   DAVIDOV, D., ORBACH, R., RETTORI, C., SHALTIEL, D., TAO, L. J., and
          RICKS, B., 1972, *Solid St. Commun.*, **10**, 451.
(A 43)   DAVIDOV, D., ORBACH, R., TAO, L. J., and CHOCK, E. P., 1971, *Physics
          Lett.* A, **34**, 379.
(A 44)   DAVIDOV, D., and SHALTIEL, D., 1968, *Phys. Rev.*, **169**, 329.
(A 45)   DAVIDOV, D., and SHALTIEL, D., 1968, *Phys. Rev. Lett.*, **21**, 1759.
(A 46)   DAVIDOV, D., RETTORI, C., BABERSCHKE, K., CHOCK, E. P., and ORBACH, R.,
          1973, *Physics Lett.* A, **45**, 161.
(A 47)   DAVIDOV, D., RETTORI, C., CHOCK, E. P., ORBACH, R., and MAPLE, M. B.,
          1972, *18th AIP Conf.*, *Denver*, **1**, 138.
(A 48)   DAVIDOV, D., MAKI, K., ORBACH, R., RETTORI, C., and CHOCK, E. P., 1973,
          Solid St. Commun., **12**, 621.
(A 49)   DAVIDOV, D., BUCHER, E., RUPP, L. W., LONGINOTTI, L. D., and RETTORI,
          C., 1974, *Phys. Rev.* B, **9**, 2879.
(A 50)   DAVIDOV, D., ORBACH, R., RETTORI, C., TAO, L. J., and RICKS, B., 1972,
          Physics Lett. A, **40**, 369.
(A 51)   DAVIDOV, D., RETTORI, C., and KIM, H. M., 1974, *Phys. Rev.*, **9**, 147.
(A 52)   DAVIDOV, D., RETTORI, C., BABERSCHKE, K., and ORBACH, R., 1973,
          Physics Lett. A, **45**, 163.
(A 53)   DAVIDOV, D., ORBACH, R., RETTORI, C., SHALTIEL, D., TAO, L. J., and
          RICKS, B., 1971, *Physics Lett.* A, **37**, 361.
(A 54)   DAVIDOV, D., RETTORI, C., ORBACH, R., DIXON, A., and CHOCK, F. P.,
          1975 (to be published).
(A 55)*   DAVIDOV, D., RETTORI, G., NG, C., and CHOCK, E. P., 1974, *Physics Lett.* A,
          49, 320.
(A 56)   DAVIDOV, D., CHELKOWSKI, A., RETTORI, C., ORBACH, R., and MAPLE,
          M. B., 1973, *Phys. Rev.*, **7**, 1029.
(A 57)   DAVIDOV, D., RETTORI, C., DIXON, A., BABERSCHKE, K., CHOCK, E. P., and
          ORBACH, R., 1973, *Phys. Rev.* B, **8**, 3563.
(A 58)*   DAVIDOV, D., RETTORI, C., and ZEVIN, V., 1975, *Solid St. Commun.*, **16**, 247.
(A 59)*   DAVIDOV, D., RETTORI, C., and SHALTIEL, D., 1974, *Physics Lett.* A, **50**, 392.
(A 60)   DEBRAY, D., and RYBA, E., 1971, *J. Phys.*, 32–C1–1132.
(A 61)   DEVINE, R. A. B., and MORET, J-M., 1974, *Solid St. Commun.*, **14**, 1287.
(A 62)   DEVINE, R. A. B., 1973, *Solid St. Commun.*, **13**, 935.
(A 63)   DEVINE, R. A. B., MORET, J. M., ORTELLI, J., SHALTIEL, D., ZINGG, W.,
          and PETER, M., 1972, *Solid St. Commun.*, **10**, 575.
(A 64)   DEVINE, R. A. B., ZINGG, W., MORET, J. M., and SHALTIEL, D., 1973, *Solid
          St. Commun.*, **12**, 515.
(A 65)   DEVINE, R. A. B., SHALTIEL, D., MORET, J. N., ORTELLI, J., ZINGG, W.,
          and PETER, M., 1972, *Solid St. Commun.*, **11**, 525.
(A 66)   DEVINE, R. A. B., FISHER, O., and ZINGG, W., 1972, *J. low temp. Phys.*,
          7, 319.
(A 67)   DEVINE, R. A. B., ZINGG, W., and MORET, J. M., 1972, *Solid St. Commun.*,
          11, 233.
(A 68)   EHARA, S., 1971, *Can. J. Phys.*, **49**, 352.
(A 69)   EHARA, S., and ARROTT, A., 1971, *17th AIP Conf.* Chicago, p. 1200.
(A 70)   ELLISTON, P. R., 1970, *Phys. Stat. Sol.* A, **9**, 111.
(A 71)   ELSCHNER, B., and WEIMANN, G., 1971, *Solid St. Commun.*, **9**, 1935.
(A 72)   ELSCHNER, B., LUFT, H., SCHÄFER, W., and SCHÖN, G., 1973, *Proc. Conf.
          on ESR of Ions in Metals* (Haute-Nendaz), p. 105.

(A 73) ENGEL, U., BABERSCHKE, K., KOOPMANN, G., HÜFNER, S., and WILHELM, M., 1973, *Solid St. Commun.*, **12**, 977.

(A 74) FISK, Z., TAYLOR, R. H., and COLES, B. R., 1971, *J. Phys. C*, **4**, L292.

(A 75) GESCHWIND, S., DEVLIN, G. E., and WERNICK, J. H., 1966, *J. appl. Phys.*, **37**, 1221.

(A 76) GOSSARD, A. C., HEEGER, A. J., and WERNICK, J. H., 1967, *J. appl. Phys.*, **38**, 1251.

(A 77) GOSSARD, A. C., KOMETANY, T. Y., and WERNICK, J. H., 1968, *J. appl. Phys.*, **39**, 849.

(A 78) GRIFFITHS, D., 1967, *Proc. phys. Soc.*, **90**, 707.

(A 79) GRIFFITHS, D., and COLES, B. R., 1966, *Phys. Rev. Lett.*, **16**, 1093.

(A 80) HACKER, H., GUPTA, R., and SHEPPARD, M. L., 1972, *Phys. Stat. Sol. A*, **9**, 601.

(A 81) HARRIS, A. M., POPPLEWELL, J., and TEBBLE, R. S., 1965, *Proc. phys. Soc.*, **85**, 513.

(A 82) HARRIS, A. M., POPPLEWELL, J., and TEBBLE, R. S., 1966, *Proc. phys. Soc.*, **88**, 679.

(A 83) HIRST, L. L., WILLIAMS, G., GRIFFITHS, D., and COLES, B. R., 1968, *J. appl. Phys.*, **39**, 844.

(A 84) HIRST, L. L., SCHÄFER, W., SEIPLER, D., and ELSCHNER, B., 1973, *Phys. Rev. B*, **8**, 64.

(A 85) JACCARINO, V., MATTHIAS, B. T., PETER, M., SUHL, H., and WERNICK, J. H., 1960, *Phys. Rev. Lett.*, **5**, 25.

(A 86) KELLER, G., LUFT, H., and ELSCHNER, B., 1974, *Physics Lett. A*, **49**, 273.

(A 87) KIP, A. F., 1953, *Rev. mod. Phys.*, **25**, 229.

(A 88) KLEINHAMS, F. W., and WIGEN, P. E., 1971, *17th AIP Conf.*, Chicago, p. 1204.

(A 89) KOJIMA, K., KASAYA, M., and KÖI, Y., 1974, *J. phys. Soc. Japan*, **36**, 1206.

(A 90) KOOPMANN, G., ENGEL, U., BABERSCHKE, K., and HÜFNER, B., 1972, *Solid St. Commun.*, **11**, 1197.

(A 91) KOOPMANN, G., ENGEL, U., HÜFNER, S., and BABERSCHKE, K., 1973, *Physics Lett. A*, **45**, 173.

(A 92)* KOOPMANN, G., BABERSCHKE, K., and HÜFNER, S., 1975, *Physics Lett. A*, **50**, 407.

(A 93) MALE, S., and TAYLOR, R. H., 1975, *J. Phys. F*, **5**, L26.

(A 94) McELROY, J. A., and HEEGER, J. A., 1968, *Phys. Rev. Lett.*, **20**, 1481.

(A 95) MILLER, D. E., and HACKER, M., JR., 1971, *Solid St. Commun.*, **9**, 881.

(A 96) MIYAKO, Y., WATANABE, T., and WATANABE, M., 1969, *Phys. Rev.*, **182**, 495.

(A 97) MIYAKO, Y., 1970, *12th Int. Conf. on Low Temp. Physics*, Kyoto.

(A 98) MORET, J. M., ORBACH, R., PETER, M., SHALTIEL, D., SUSS, J. T., ZINGG, W., and DEVINE, R. A. B., 1975, *Phys. Rev.*, **11**, 2002.

(A 99) NAGASHIMA, H., and ABE, H., 1971, Technical Report, ISSP, 489A.

(A 100) NAGEL, J., HÜFNER, S., and GRÜNIG, M., 1973, *Solid St. Commun.*, **8**, 1279.

(A 101) NAKAMURA, A., and KINOSHITA, N., 1967, *J. phys. Soc. Japan*, **22**, 335.

(A 102) NAKAMURA, A., and KINOSHITA, N., 1967, *J. phys. Soc. Japan*, **23**, 449.

(A 103) NAKAMURA, A., and KINOSHITA, N., 1969, *J. phys. Soc. Japan*, **26**, 48.

(A 104) NAKAMURA, A., NISHIHARA, Y., and KINOSHITA, N., 1970, *Proc. 12th Int. Conf. on Low Temp. Phys.*, Kyoto.

(A 105) ODA, Y., and ASSAYAMA, K., 1968, *J. phys. Soc. Japan*, **25**, 1510.

(A 106) ODA, Y., and ASAYAMA, K., 1970, *J. phys. Soc. Japan*, **29**, 869.

(A 107) OKUDA, K., and DATE, M., 1967, *J. phys. Soc. Japan*, **22**, 1512.

(A 108) OKUDA, K., and DATE, M., 1968, *J. phys. Soc. Japan*, **25**, 1732.

(A 109) OKUDA, K., and DATE, M., 1969, *J. phys. Soc. Japan*, **27**, 839.

(A 110) OWEN, J., BROWNE, M. E., ARP, V., and KIP, A. F., 1957, *J. Phys. Chem. Solids*, **2**, 85.

(A 111) OWEN, J., BROWNE, M. E., KNIGHT, W. D., and KITTEL, C., 1956, *Phys. Rev.*, **102**, 1501.

(A 112) PETER, M., 1961, *J. appl. Phys.*, **32**, S338.

(A 113) PETER, M., and MATTHIAS, B. T., 1960, *Phys. Rev. Lett.*, **4**, 449.

(A 114)   PETER, M., SHALTIEL, D., WERNICK, J. H., WILLIAMS, M. J., MOCK, J. B.,
          and SHERWOOD, R. C., 1962, *Phys. Rev. Lett.*, **9**, 50.
(A 115)   PETER, M., SHALTIEL, D., WERNICK, J. H., WILLIAMS, H. J., MOCK, J. B.,
          and SHERWOOD, R. C., 1962, *Phys. Rev.*, **126**, 1395.
(A 116)   PETER, M., SHALTIEL, D., WERNICK, J. H., WILLIAMS, H. J., MOCK, J. B.,
          and SHERWOOD, R. C., 1968, *Phys. Rev.*, **126**, 1395.
(A 117)   POP, I., and CHECHERNIKOV, V. I., 1965, *Soviet Phys. solid St.*, **6**, 3011.
(A 118)   POPPLEWELL, J., and TEBBLE, R. S., 1963, *J. appl. Phys.*, **34**, 1343.
(A 119)   PRADDAUDE, M. C., 1972, *Physics Lett.* A, **42**, 97.
(A 120)   PRADDAUDE, M. C., GUERTIN, R. P., FONER. S., and McNIFF, E. J., 1972,
          18th Mag. Conf., AIP, Denver, p. 1115.
(A 121)   PRADDAUDE, M. C., and GÄRTNER, H., 1971, *Physics Lett.* A, **34**, 217.
(A 122)   RETTORI, C., DAVIDOV, D., ORBACH, R., CHOCK, E. P., and RICKS, B., 1973,
          Phys. Rev. B, **7**, 1.
(A 123)   RETTORI, C., DAVIDOV, D., CHAIKIN, P., and ORBACH, R., 1973, *Phys. Rev.
          Lett.*, **30**, 437.
(A 124)   RETTORI, C., KIM, H. M., CHOCK, E. P., and DAVIDOV, D., 1974, *Phys. Rev.*,
          10, 1826.
(A 125)* RETTORI, C., and DAVIDOV, D., 1974, *Phys. Rev.*, **10**, 4033.
(A 126)   RODBELL, D. S., and MOORE, T. W., 1963, *Proc. Int. Conf. Magn.*, Notting-
          ham, pp. 427.
(A 127)   SALAMON, M. B., 1971, *Phys. Rev. Lett.*, **26**, 704.
(A 128)   SALAMON, M. B., and FEIGL, F. J., 1968, *J. Phys. Chem. Solids*, **29**, 1443.
(A 129)   SCHÄFER, W., SCHMIDT, H. K., and HUFNER, S., 1968, *Physics Lett.* A, **28**,
          279.
(A 130)   SCHÄFER, W., SCHMIDT, H. K., ELSCHNER, B., and BUSCHOW, K. H. J.,
          1970, *Physics Lett.* A, **33**, 23.
(A 131)   SCHÄFER, W., SCHMIDT, H. K., HÜFNER, S., and WERNICK, J. H., 1969,
          Phys. Rev., **182**, 459.
(A 132)   SCHÄFER, W., SCHMIDT, H. K., ELSCHNER, B., and BUSCHOW, K. H. J.,
          1972, *Z. Phys.*, **254**, 1.
(A 133)   SCHMIDT, H. K., 1972, *Z. Naturf.* A, **27**, 191.
(A 134)   SCHMIDT, H. K., SCHÄFER, W., KELLER, G., and ELSCHNER, B., 1972,
          Physics Lett. A, **38**, 201.
(A 135)* SCHRITTENLACHER, W., BABERSCHKE, K., KOOPMANN, G., and HÜFNER, S.,
          1975, *Solid St. Commun.* (to be published).
(A 136)   SHALTIEL, D., 1963, *J. appl. Phys.*, **34**, 1190.
(A 137)   SHALTIEL, D., GOSSARD, A. C., and WERNICK, J. H., 1965, *Phys. Rev.*, **137**,
          A1027.
(A 138)   SHALTIEL, D., and WERNICK, J. H., 1964, *Phys. Rev.*, **136**, A245.
(A 139)   SHALTIEL, D., WERNICK, J. H., and JACCARINO, V., 1964, *J. appl. Phys.*,
          35, 978.
(A 140)   SHALTIEL, D., WERNICK, J. H., WILLIAMS, H. J., and PETER, M., 1964,
          Phys. Rev., **135**, A1346.
(A 141)   SPERLICH, G., and JANNECK, K. H., 1972, *Int. J. Magn.*, **3**, 157.
(A 142)   SPERLICH, G., JANNECK, K. H., and BUSCHOW, K. H. J., 1973, *Phys. Stat.
          Sol.* B, **47**, 701.
(A 143)   SPERLICH, G., 1973, *Proc. Conf. on ESR of ions in Metals* (Haute-Nendaz),
          p. 117.
(A 144)   STREET, R., 1960, *J. appl. Phys.*, **31**, 310.
(A 145)   SUN, S. F., and SCHNITZKE, K., 1971, *Physics Lett.* A, **35**, 263.
(A 146)   TAO, L. J., DAVIDOV, D., ORBACH, R., and CHOCK, E. P., 1971, *Phys. Rev.* B,
          4, 5.
(A 147)   TAO, L. J., DAVIDOV, D., ORBACH, R., SHALTIEL, D., and BURR, C. R., 1971,
          Phys. Rev. Lett., **26**, 1438.
(A 148)   TAYLOR, R. H., and COLES, B. R., 1973, *Proc. Conf. on ESR of Ions in
          metals* (Haute-Nendaz), p. 99.
(A 149)   TAYLOR, R. H., 1973, *J. Phys.* F, **3**, L110.
(A 150)   TAYLOR, R. H., and COLES, B. R., 1974, *J. Phys.* F, **4**, 303.

(A 151)　TAYLOR, R. H., and COLES, B. R., 1975, *J. Phys.* F, **5**, 121.
(A 152)*　URBAN, P., ELSCHNER, B., SPERLICH, G., and SCHÖN, G., 1974, *Physics Lett.* A, **49**, 418.
(A 153)　URSU, I., and BURZO, E., 1972, *J. magn. Reson.*, **8**, 274.
(A 154)　VIJARAGHAVAN, R., RAO, V. U. S., and MALIK, S. K., 1968, *J. appl. Phys.*, **39**, 1086.
(A 155)　VIJARAGHAVAN, R., MALIK, S. K., and RAO, V. U. S., 1968, *Phys. Rev. Lett.*, **20**, 106.
(A 156)　WEIMANN, G., BOPP, P. M., ELSCHNER, B., and BUSCHOW, K. H. J., 1973, *Int. J. Magn.*, **5**, 1.
(A 157)　WEIMANN, G., ELSCHNER, B., BUSCHOW, K. H. J., and VAN STAPELE, R. P., 1972, *Solid St. Commun.*, **11**, 871.
(A 158)　WEIMANN, G., and ELSCHNER, B., 1973, *Z. Phys.*, **261**, 85.
(A 159)　WERNICK, J. H., WILLIAMS, H. J., and GOSSARD, A. C., 1967, *J. Phys. Chem. Solids*, **28**, 271.
(A 160)　ZINGG, W., BILL, H., BUTTET, J., and PETER, M., 1974, *Phys. Rev. Lett.*, **32**, 1221.
(A 161)　ARBILLY, D., DEUTSCHER, G., GRÜNBAUM, E., ORBACH, R., and SUSS, J. T., 1975, *Phys. Rev.* B, **12**, 5068.
(A 162)　BAJGULEY, D. M. S., and PARTINGTON, J. P., 1975, *J. Phys.* F, **5**, L39.
(A 163)　BHAGAT, S. M., and PAUL, D. K., 1975, *Phys. Rev. Lett.*, **35**, 1458.
(A 164)　DAVIDOV, D., and BABERSCHKE, K., 1975, *Phys. Rev. Lett.*, A**151**, 144.
(A 165)　DAVIDOV, D., ZEVIN, V., BLOCH, J. M., and RETTORI, C., 1975, *Solid St. Commun.*, **17**, 1279.
(A 166)　DEVINE, R. A. B., CHIU, J. C. M., POIRIER, M., 1975, *J. Phys.* F, **5**, 2362.
(A 167)　DEVINE, R. A. B., POIRIER, M., and CYR, T., 1975 *J. Phys.* F, **5**, 1407.
(A 168)　KLEINHAMS, F. W., LONG, J. P., and WIGEN, P. E., 1975, *Phys. Rev.* B, **11**, 2638.
(A 169)　MALIK, S. K., and VIJARAGHAVAN, R., 1975, *Phys. Rev.* B, **12**, 3971.
(A 170)　OSEROFF, S., GEHMAN, B., SCHULTZ, S., and RETTORI, C., 1975, *Phys. Rev. Lett.*, **35**, 679.
(A 171)　PRADDAUDE, H. C., 1975, *Solid St. Commun.*, **16**, 1019.
(A 172)　RETTORI, C., DAVIDOV, D., GRAYEVSKEY, A., and WALSH, W. M., 1975, *Phys. Rev.* **11**, 4450.
(A 173)　RETTORI, C., DAVIDOV, D., NG, G., and CHOCK E., P., 1975, *Phys. Rev.* B, **12**, 1298.
(A 174)　RETTORI, C., LEVIN, R., and DAVIDOV, D., 1975, *J. Phys.* F, **5**, 2379.
(A 175)　SEIPLER, D., and ELSCHNER, B., 1975, *Physics Lett.* A, **55**, 115.
(A 176)　SJÖSTRAND, M. E., and SEIDEL, G., 1975, *Phys. Rev.* B, **11**, 3292.
(A 177)　URBAN, P., DAVIDOV, D., ELSCHNER, B., PLEFKA, T., and SPERLICH, G., 1975, *Phys. Rev.* B, **12**, 72.
(A 178)　ZINGG, W., BUTTET, J., and HARDIMAN, M., 1975, *Proc. 18th Coll. Ampére*, Nottingham, Vol. 2 (North-Holland), pp. 449–50.

Author Index

Lea, M. R., 15, 24, 88
Liu, S. H., 5
Longo, R. T., 73

Maki, K., 43, 44, 45
Male, S., 14, 37
Matthias, B. T., 26
McElroy, J. A., 71
Miller, D. E., 7, 30
Miyako, Y., 72, 76
Monod, P., 1
Moore, T. W., 26
Moret, J. M., 17, 49, 52, 77, 92
Murani, A. P., 80

Nagashima, H., 72
Nagel, J., 22, 90
Nakamura, A., 25, 62, 71
Narath, A., 9

Oda, Y., 71
Okuda, K., 8, 27, 71
Orbach, R., 1, 12, 16, 17, 67, 71, 81, 93
Overhauser, A., 12
Owen, J., 20, 68, 69, 70, 72, 76

Peter, M., 1, 6, 8, 12, 26, 46, 47, 48, 49, 51, 65
Phillips, W. C., 6
Pifer, J. H., 73
Plefka, T., 17, 56
Pop, I., 55, 60
Popplewell, J., 26, 54
Praddaude, M. C., 49, 52, 84
Purwins, H. G., 88

Rettori, C., 20, 21, 22, 39, 40, 42, 43, 44, 64, 67, 85
Rivier, N., 20
Rizzutto, C., 22
Rodbell, D. S., 26
Ryba, E., 8, 29

Salamon, M. B., 12, 54, 59, 67
Sarkissian, B., 58
Schäfer, W., 37, 55
Schaller, H. J., 49
Schmidt, H. K., 12, 14, 55, 59
Schnitzke, K., 42
Schneffer, J. R., 5
Schultz, S., 1, 73
Shaltiel, D., 6, 7, 8, 14, 20, 26, 27, 28, 34, 35, 36, 46, 47, 49, 51, 69, 73, 75
Shull, C. G., 6
Spencer, H. J., 12, 22
Sperlich, G., 7, 31, 32, 41, 42
Stewart, A. M., 10
Street, R., 69
Switendick, A. C., 44
Sun, S. F., 42

Tao, L. J., 18, 54, 63, 64, 67, 80
Taylor, K. N. R., 33
Taylor, R. H., 7, 8, 14, 19, 24, 27, 28, 31, 32, 33, 37, 39, 49, 52, 77
Tebble, R. S., 26, 54

Ursu, I., 8, 10, 28

Van Diepen, A. M., 10
Vijaraghavan, R., 8, 27

Walker, M. B., 22
Watson, R. E., 6, 10
Weaver, H. R., 9
Weimann, G., 7, 28, 41, 56, 57, 61
Weiss, P. R., 32
Wernick, J. H., 7, 27, 69, 73, 75
Wigen, P. E., 76
Williams, G., 16, 63, 82, 89
Wolff, P. A., 5

Yafet, Y., 12, 14
Yamada, Y., 6
Yosida, K., 4

Zimmermann, P. H., 17, 49
Zingg, W., 83, 88, 92

Subject Index

The reader is also referred to Part 1 of the Appendix (pages 97-106) and to its Addendum (pages 107-108).   In the last column on those pages, the numbers refer to the section headings of Part 2.